THE MOON

NASA IMAGES FROM SPACE

Beth Alesse

Amherst Media, Inc. ■ Buffalo, NY

Beth Alesse is an author, graphic artist, and editor. She curates image collections to present in books and media. She holds degrees in art and education, and has backgrounds in graphic arts, linguistics, and visual and digital media. In addition to *The Moon,* her books in the Amherst Media space series include *The Earth: A Visual Story of Our Amazing Planet*, *The Sun: NASA Images from Space* and *Hubble in Space: NASA Images of Planets, Stars, Galaxies, Nebulae, Black Holes, Dark Matter, & More.*

Published by:
Amherst Media, Inc., P.O. Box 538, Buffalo, N.Y. 14213
www.AmherstMedia.com

Publisher: Craig Alesse
Associate Publisher: Katie Kiss
Senior Editor/Production Manager: Michelle Perkins
Editors: Barbara A. Lynch-Johnt, Beth Alesse
Acquisitions Editor: Harvey Goldstein
Editorial Assistance from: Ray Bakos, Carey Miller, Rebecca Rudell, Jen Sexton-Riley
Business Manager: Sarah Loder
Marketing Associate: Tonya Flickinger

ISBN-13: 978-1-68203-368-5
Library of Congress Control Number: 2018936016
Printed in The United States of America.
10 9 8 7 6 5 4 3 2 1

www.facebook.com/AmherstMediaInc
www.youtube.com/AmherstMedia
www.twitter.com/AmherstMedia
www.instagram.com/amherstmediaphotobooks

Contents

Introduction

Legends from ancient cultures are full of stories in which the Sun and the Moon were deified. In their godly forms the spheres acknowledge each other but keep to their own sectors of the sky, and quite alone. For most of human history, the Moon has been a solitary orbit, unreachable and untouchable in the sky.

The relationship between humans and the Moon changed with the development of telescopes. These instruments gave us the means to focus on the Moon's details. Theories about the Moon and the celestial bodies advanced.

The Apollo astronauts stepping on the Moon changed our thinking forever. Humankind's potential could be seen in a different light. Our arrival on the Moon changed and continues to affect how we think and view science, business, art, and politics. The Moon is central to the Earth's past and future. Today the Moon is like the unexplored continents of long ago. We are slowly mapping and investigating it. Finding its scientific treasures—its useful resources—will help us to exist in this off-world land and one day other distant celestial-worlds.

This book includes many images from NASA, its collaborators, and other sources. Some images are historical and in the public domain. Consideration has been given to properly credit images or place them in their correct historical, public domain context. Please contact with any corrections for future editions at:

BethAlesse@AmherstMedia.com

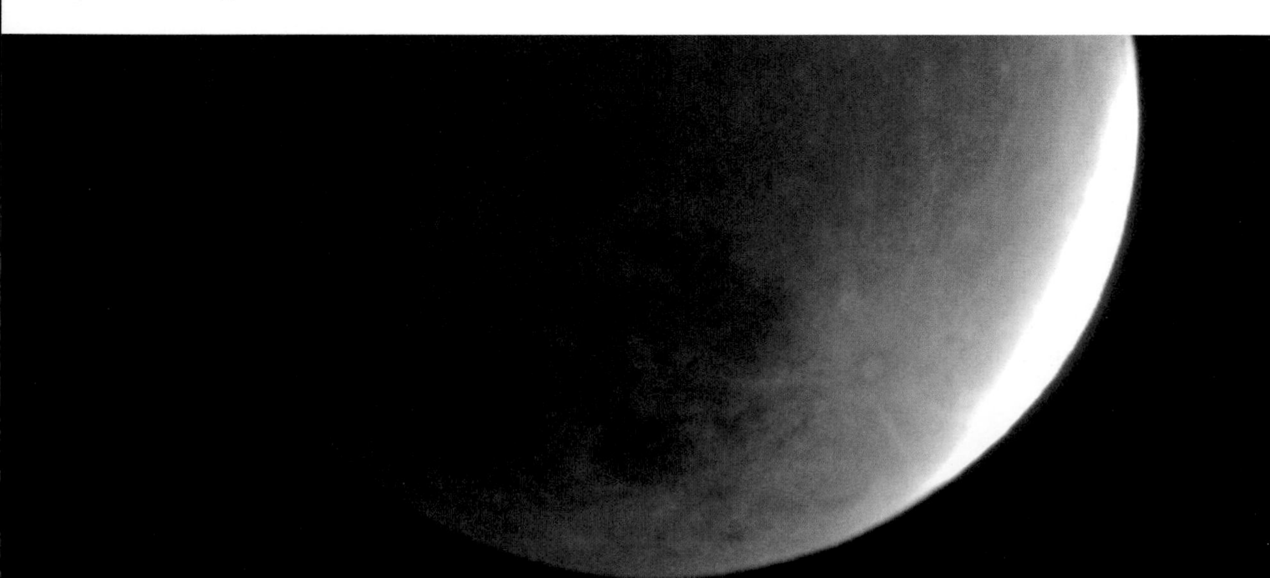

Moon Basics

Image credit: NASA/ Bill Ingalls

Some Lunar Facts

The Moon is Earth's single, natural satellite. There are 194 natural satellites in the Solar System. There are four larger natural planetary satellites in the Solar System belonging to the planets Jupiter and Saturn.

The Moon is thought to be nearly as old as Earth, about 4.5 billion years old. From its center outward—its radius, it is about 1,079 miles (1738 kilometers), which is about one fourth that of the Earth's radius. This makes Earth the planet with the largest satellite-to-planet-size ratio.

Its equatorial circumference is about 6785, compared to the Earth's circumference of approximately 24,900. The Moon's distance to Earth is 238,900 miles.

"By the Light of the Silvery Moon"

From the beginning of humanity's existence to the present day, the Moon has added to our ability to see at night, periodically lengthening our active hours. Animals also use it for navigation. The Moon's light that we see in the night sky and sometimes during the day is not made by the Moon itself. It is light from the Sun reflected by the Moon's surface.

Variations in the Sky

From the Earth, the Moon can appear differently: a crescent, a disk, white, yellow, small, voluminous, high, low, in the night, and in the day, crisp or hazy, clear or shrouded. The left image was taken over Bangladesh in 2016 from a jet plane en route to Hanoi, Vietnam with U.S. Secretary of State John Kerry . The red moon image *(right)* is of the Moon as it was covered by the umbra during an eclipse.

Image credit *(right):* NASA
Image credit *(left):* State Department

Image credit: NASA

Movement Across the Sky

Standing on the Earth's surface, the Moon, just like the Sun, appears to move daily across the sky, rising in the east and setting in the west. This apparent daily movement is a result of the Earth's rotation and not an actual movement by the Moon or Sun.

Periodic Movement Around the Earth

The Moon revolves around the Earth taking approximately a month to complete its trip. As the Moon goes through its monthly cycle, its appearance changes. This is referred to as phases. The phases are described mainly in terms of their shape: new, quarter, gibbous, full, and crescent. Changes in the Moon's shape happen because, from our perspective on Earth, these are parts of the Moon's sphere that are illuminated by the Sun and not in shadow.

Apparent Movements

The daily Moon movement appears to advance across the nightly sky in the same direction as does the Sun, in a daily east to west course. A full Moon during the spring or autumn equinox appears to rise when the Sun sets and sets when the Sun rises. This timing varies with the shortening and lengthening of the day throughout the year.

The southerly-northerly degree of the daily arc changes with the Earth's seasons. This is a result of the Earth's tilt with respect to the Sun. Other apparent movements of the Moon can be detected against the celestial sphere. The shifting location of objects within the sphere's backdrop is partially a consequence of the Earth-Moon system's yearly trek around the Sun.

Synchronous Rotation

The same side of the Moon always faces Earth. Its near side faces us. Looking up from Earth, we will never see the far side. This is becausc the Moon is tidally locked, referred to as in a synchronous rotation. As a result, one day for the Moon lasts a complete month—sunrise to sunrise. One lunar day equals about 27 Earth days.

The near side that faces Earth has shown its face from humankind's beginning. The far side of the Moon was not seen or documented until the Soviet probe Luna 3 took the very first photographs.

Lunar Orbital Periods

An orbital period is the measure of time it takes for one astronomical object to orbit around another. In general, we call this period of the Moon's orbit a month. It's orbit period is calculated two ways. The sidereal month of 27.3 days is the true orbital period. However from our perspective on the Earth, the synodic month takes 29.5 days for the Moon to return to the same place within the celestial sphere—from new moon to new moon. Because the Earth is also in orbit around the Sun, realignment takes another 2.2 days more than the actual orbit period for a visually apparent new moon.

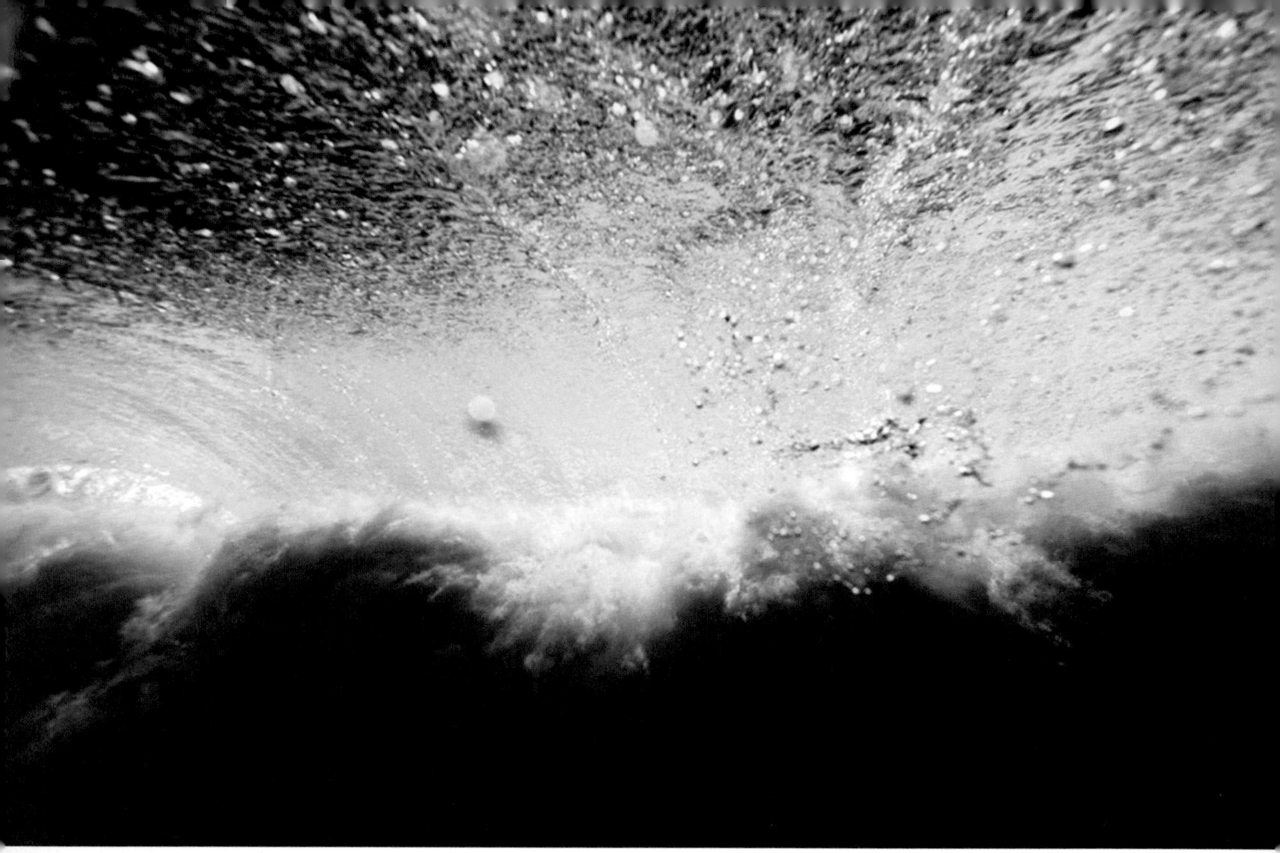

Image credit: US Department of Energy

Tidal Forces

The Moon is tidally locked to the Earth and has a synchronous rotation resulting in its lengthy lunar day. Even though the Earth is considerably larger than its satellite, the Moon, it nonetheless has tidal affects on the planet's crust. The solid part of the Earth deforms slightly. However, its affects on the ocean waters is significant. The sea levels regularly rise and fall by the joint gravitational force of the Moon and Sun on the rotating Earth. In addition to the gravitational pull from large objects in space, many things on Earth can affect the timing of tides in different localities (for example, weather conditions in the atmosphere and on the surface). Also, bathymetrics—the arrangement and depth of features under water—has a strong affect on tides.

Tidal forces are bidirectional in the case of the Earth and Moon. The Earth's gravitational pull on the Moon is even stronger than the Moon's pull on the Earth because of its greater mass.

Other Moons

There are 24 moons around planets in our Solar system. Dwarf planet Pluto and its moon Charon are classified by some as a binary system because the center of their orbits is not inside either of their bodies. Earth's satellite, the Moon, is not the largest moon in the solar system, but it is the largest in relation to its planet's size. Each of the planet-moon systems is unique.

Image credit: NASA

Scale of Other Moons in the Solar System

The illustration below shows the scales of the moons of our planetary system in relation to the size of the Earth. The composition of the moons varies, some containing water and an atmosphere. Others, like Mars' Phobos and Deimos, are quite small, but may nonetheless be key in our interplanetary explorations.

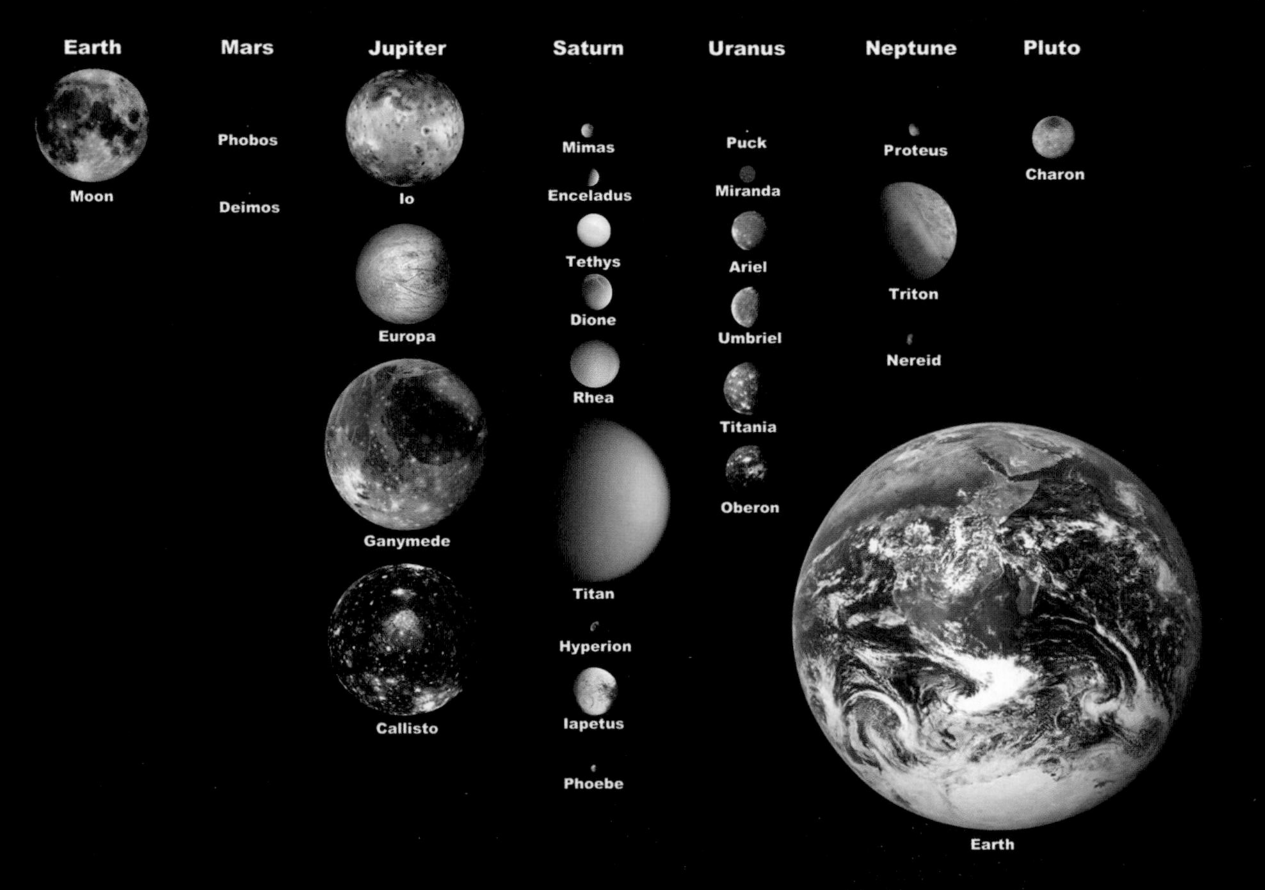

Image credit: NASA/JPL-Caltech

Collision with a Planet Called Theia

It is theorized that a planet about the size of Mars called Theia impacted an early version of the Earth around 4.5 billion years ago. Theia in Greek mythology is the mother of Selene (the Moon). Earth is often referred to as Proto-Earth at this time of its infancy.

There are variations of this giant-impact hypothesis, each one trying to better account for the data and evidence that exists today.

More Moon Facts

- The Moon is the fifth largest of all the known moons in the Solar System.
- Next to the Sun, it is the brightest object in our sky.
- It is the brightest object in the night sky.
- The Earth's axis wobble is stabilized by the Moon.
- The Moon, along with the Sun causes tides in the oceans on Earth.
- The Moon has been a timepiece for humans on which the month is based.
- The Moon provides evening light, a source of direction, and timing for life on Earth.
- The Moon is the only other celestial body besides Earth that humans have visited.
- The Moon used to be closer to the Earth, and it moves about an inch farther away every year.
- The Moon's gravity is 16.6 percent that of the Earth's gravity.
- The Moon was formed 4.6 billion years ago and about 30–50 million years after the solar system's formation.
- The Moon's rotation is synchronous. The same side always faces the Earth.
- The average distance to the Moon is 238,855 miles (384,400 kilometers) away from the Earth. About 30 Earths could fit in this distance.
- The Moon has a solid iron-rich core that is surrounded by a liquid iron layer.
- The Moon also has a mantle and a crust.

Image credits *(facing page):* NASA/JPL-Caltech

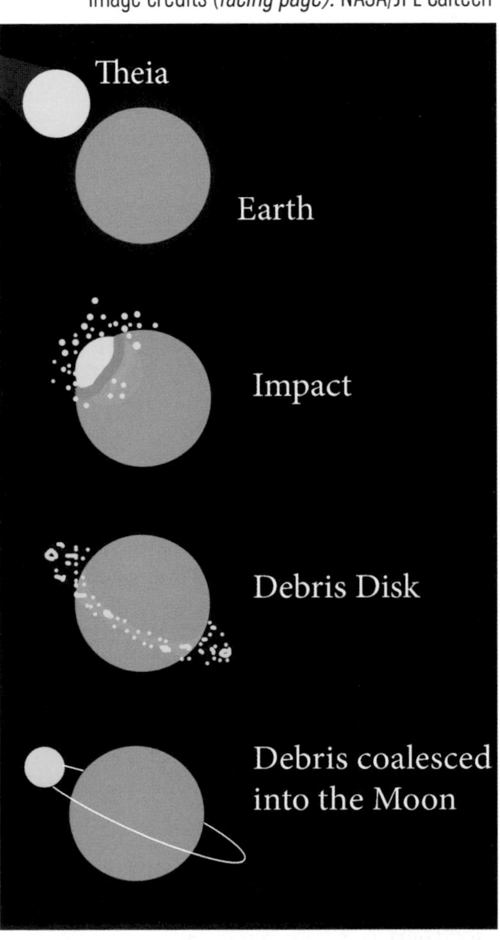

Appearance of Moons

Through Atmosphere

The Moon can look very different from one day to another or from one minute to another, or disappear from view altogether. The atmosphere where the viewer stands is one variable. The humidity, cloud coverage, the time of day, and pollution can influence visibility of the Moon's disc in the sky and the distinction of its details. Weather most often obscures our view. It is also possible for a moon to have atmospheric conditions that affect the clarity of viewing.

Earth's Rotation

The Earth's rotation affects the appearance and location of the Moon in our skies. Like the Sun, the Moon rises and sets. The amount of atmosphere the Moon's reflected light must travel through to reach our eyes is greater at moonrise and set than when it is at its apex. Consequently, the greater atmosphere acts to magnify the disc, making it appear larger than it is near the horizon.

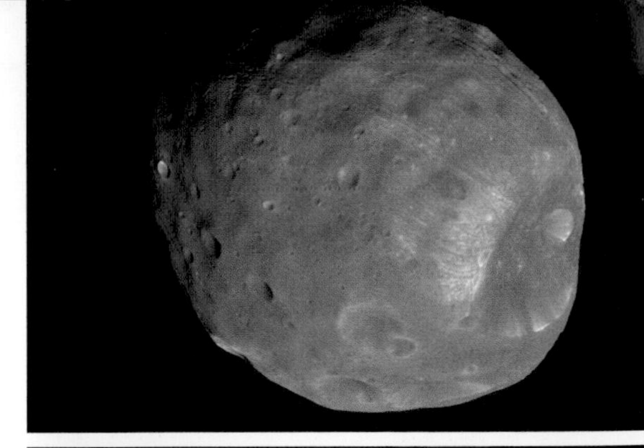

Configuration of Celestial Bodies

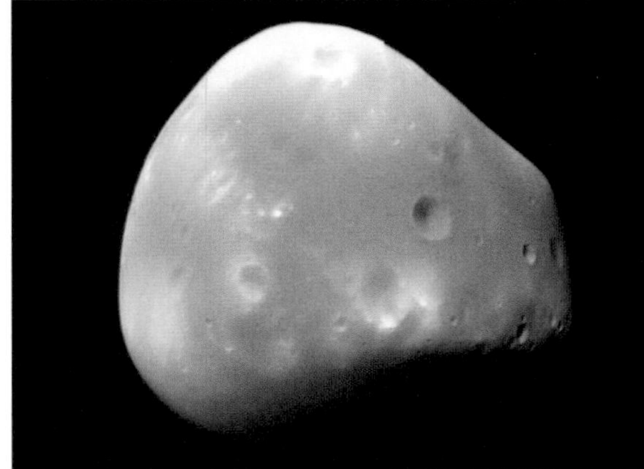

The shape of the Moon's disc changes in phases. The Moon is a sphere that appears as a somewhat two dimensional disc in our sky. Since the Moon travels around the Earth, our relative viewing position in relation to its lighting, also changes. That is why we see the moon going through phases.

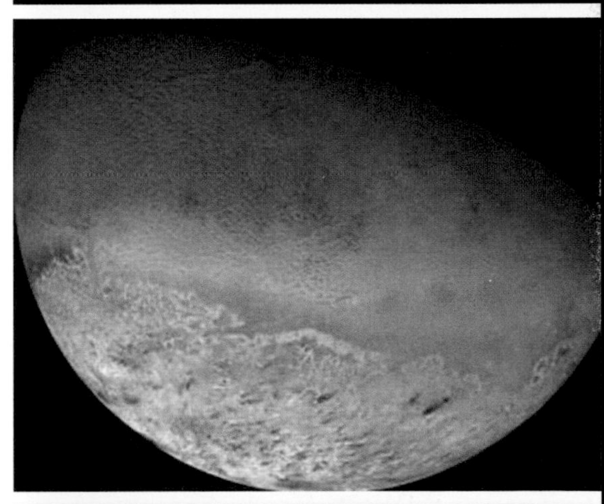

Spherical or Not

Larger moons *(facing page)* tend to be more spherical. Smaller moons *(right)* will be more irregular in shape. This is because more massive bodies have a greater amount of gravity. Higher amounts of gravity will pull the object into a sphere. Whereas the smaller body doesn't have the gravitational force to accomplish this.

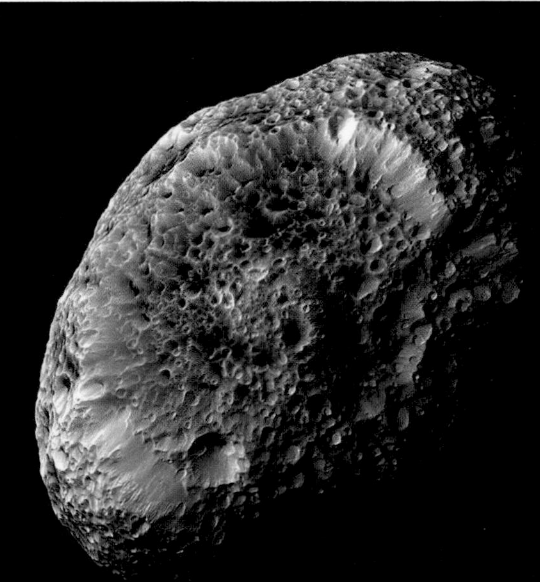

Image set credit: NASA

Image set credit: NASA-

Why the Moon Looks as Big as the Sun

The Moon is smaller than the Earth and the Sun. Its diameter is approximately 25 percent of the Earth's diameter. The diameter of thc Sun is about 400 Moons across. So why do the Sun and Moon look to be the same size when seen in the sky from Earth's surface? During a solar eclipse, it appears large enough to cover the Sun and create a shadow on the Earth's terrestrial surface.

Linear perspective can help explain the apparent size of the Sun and Moon. The closer an object is to the viewer, the larger the object appears. Consider a line of poles alongside a road. As the road and poles recede in the distance, they appear smaller. We know the distant poles are not smaller, just their apparent size on the picture plane is.

The Moon is approximately 238,900 miles (384,400 km) from Earth and is relatively close to the Earth's surface, much closer than the Sun, which is approximately 93,000,000 miles/km. The Sun is so far away, it appears much smaller than its 864,938 mile (1.392 million km) diameter.

In the past the Moon was even closer to the Earth. At that time, it would have an apparent size larger than the Sun. The Moon is slowly moving farther from the Earth, and as a result it will appear smaller. Eventually, it will not be close enough to completely cover the Sun during a solar eclipse.

Neptune's Crescent and Its Crescent Moon

The planet Neptune and its moon Triton wouldn't show a crescent from our terrestrial vantage. However, this image was taken by the Voyager 2 spacecraft in 1989 before visiting Uranus, Saturn, and Jupiter. Voyager 2 captured Neptune's crescent and a mirroring crescent in Triton.

The Phases of the Moon

Generally, the appearance of the disks of spherical objects in space changes with regularity as they move in their orbits. Different amounts of an object's disc are seen. As an orbiting object is lit by a star, light is reflected in the direction of the observer. From Earth, we observe the Moon in its different phases, and the source of the Moon's reflected light is the Sun. However, when we receive recorded images from space, the vantage point could be from the Moon, an artificial satellite, a spacecraft, or another planet. If the direction and distance of the light, the object reflecting the light, and viewpoint of the observer or camera is right, the disc of the spherical object can appear in different phases. Venus' phases can be seen with a simple telescope from the Earth's surface.

Full phase is a fully lit disc. The disc is reducing with a waning gibbous, third quarter, and waning crescent. A new moon is not visible or can sometimes be seen as a faint shadow in the daytime sky. The reason why the Moon can't be seen is because it is directly between the Earth and the Sun. At this time the Moon reflects very little light in the Earth's direction, and what little light there is, is washed out in the bright daylight. The Moon's disc increases in size after its new phase with waxing crescent, first quarter, and waxing gibbous phases, until the Moon is full again. It takes the Moon about 29.5 days (called the sidereal month) to go through the cycle of its phases. This is a little longer than what it actually takes—27 days, 7 hours, and 43 minutes.

Image credits: NASA/MSFC/Debra Needham; Lunar and Planetary Science Institute/David Kring

Image set credit: NASA

The Moon's Umbra

Although solar eclipses happen every six months or so, the path through any particular country or region could be a once-in-a-lifetime occurrence for viewers.

The umbra, where the Moon's shadow is darkest, is quite small. The penumbra, the partial shadow, encompasses a much larger area on the Earth's surface, creating a partial eclipse.

Image credit: NASA's Scientific Visualization Studio

Image set credit: NASA

Solar Eclipse 2017

A solar eclipse happens when the Moon blocks the Sun's light from reaching the Earth. The light-obscured surface falls into a shadow.

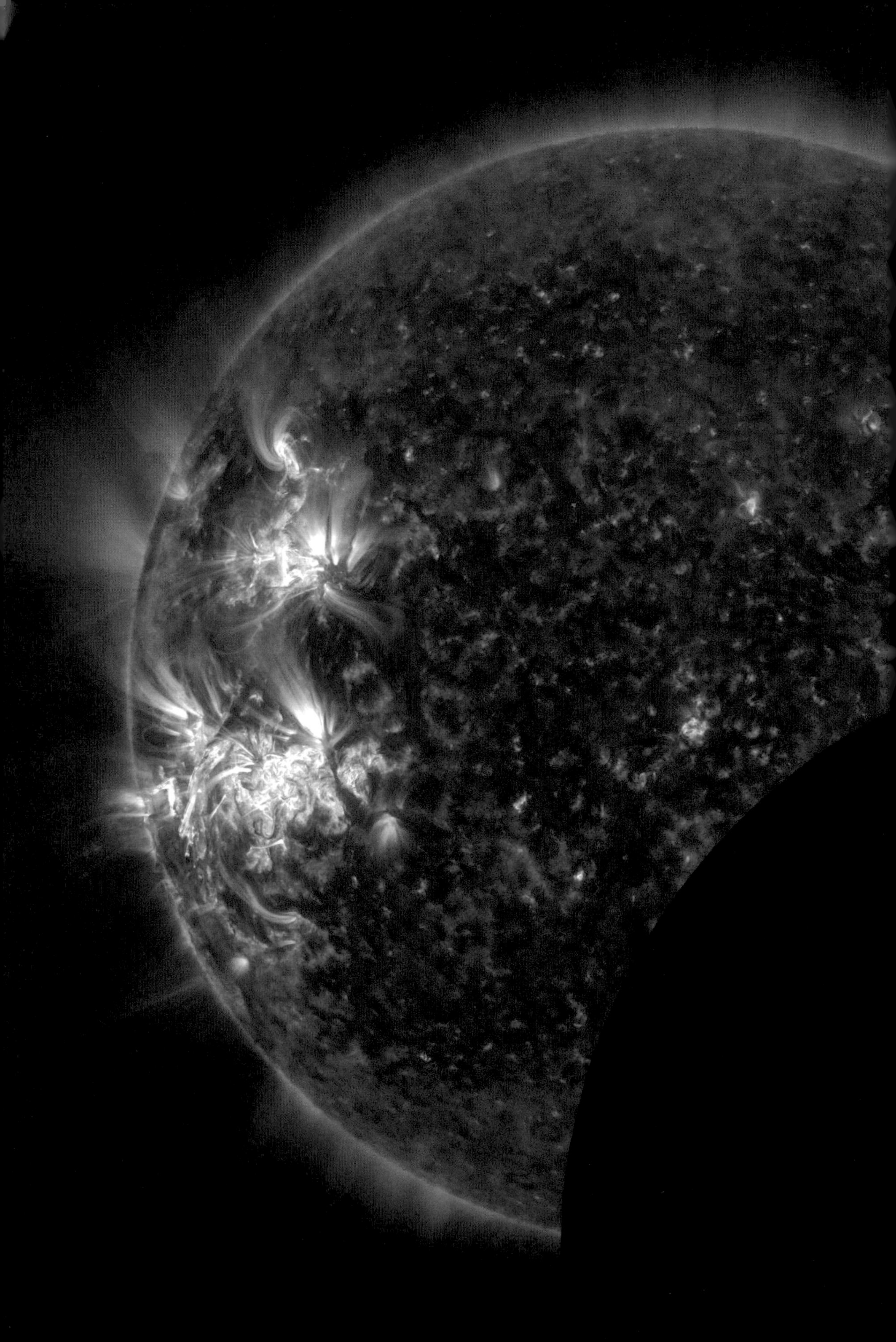

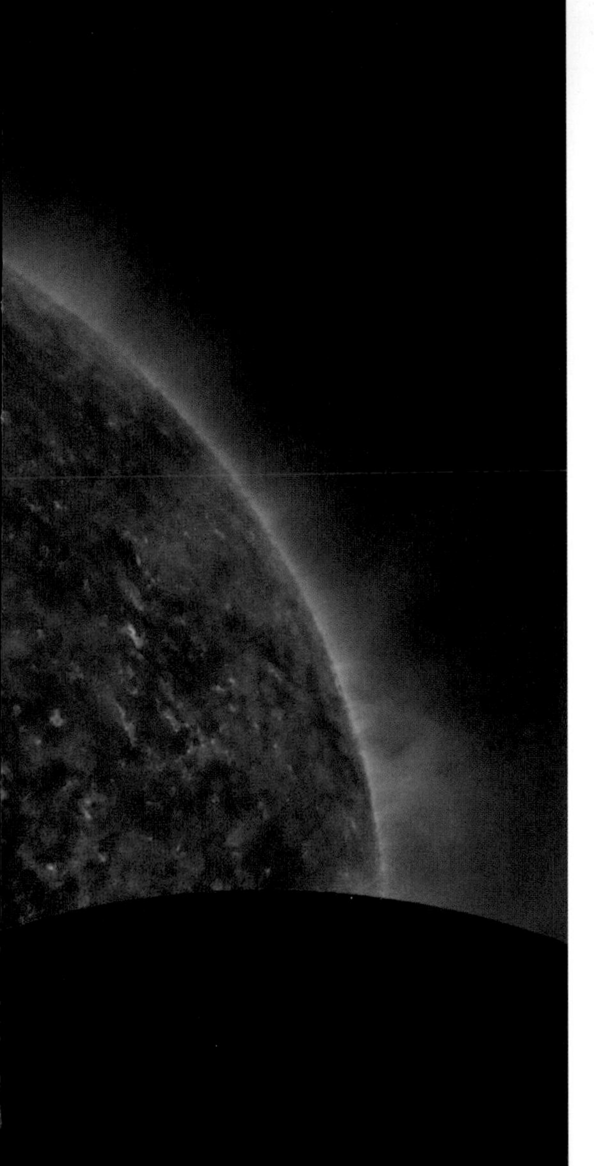

Image credit: NASA

Safe Viewing

It is not safe to look into the Sun. It will harm your eyes. Common sense tells us not to look at the Sun, and this goes for viewing a solar eclipse. Keep these points in mind:

- A solar eclipse totality, when the Sun is completely covered by the Moon, lasts only a few minutes.
- Use solar filters and viewers from a reputable vendor.
- Do not use solar filters or viewers if scratched, damaged, or more than a few years old.
- Supervise children.
- Do not use the naked eye, binoculars, telescopes, sunglasses, films, or most filters.
- Filters and viewers should not be used if incorrect grade, scratched, faulty, or attached improperly.

This image was taken by an instrument designed to record the sun. Safe viewing options include online viewing, creating a shadow of the eclipse, meeting up with astronomy groups, and using correctly made and labeled protective lenses.

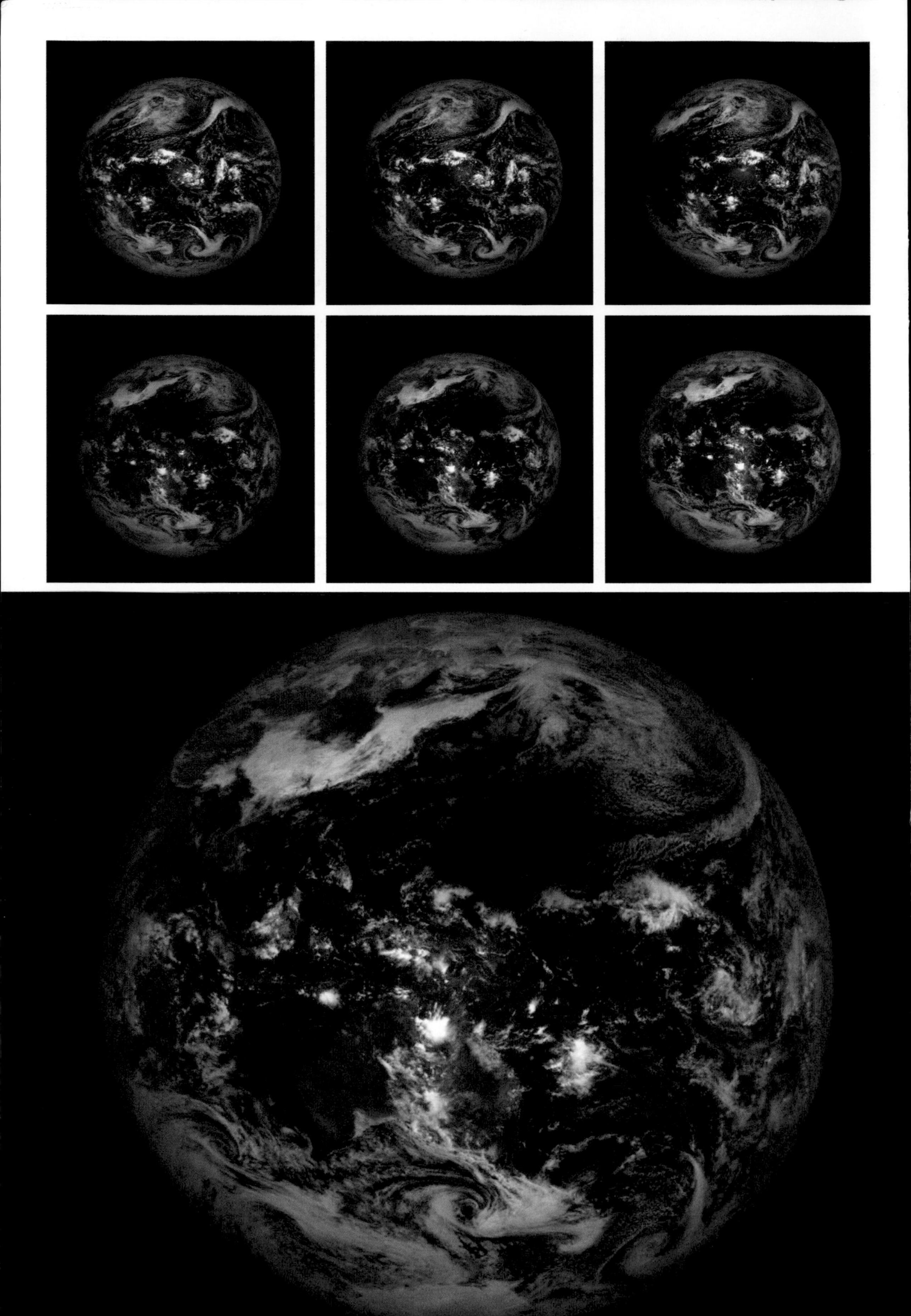

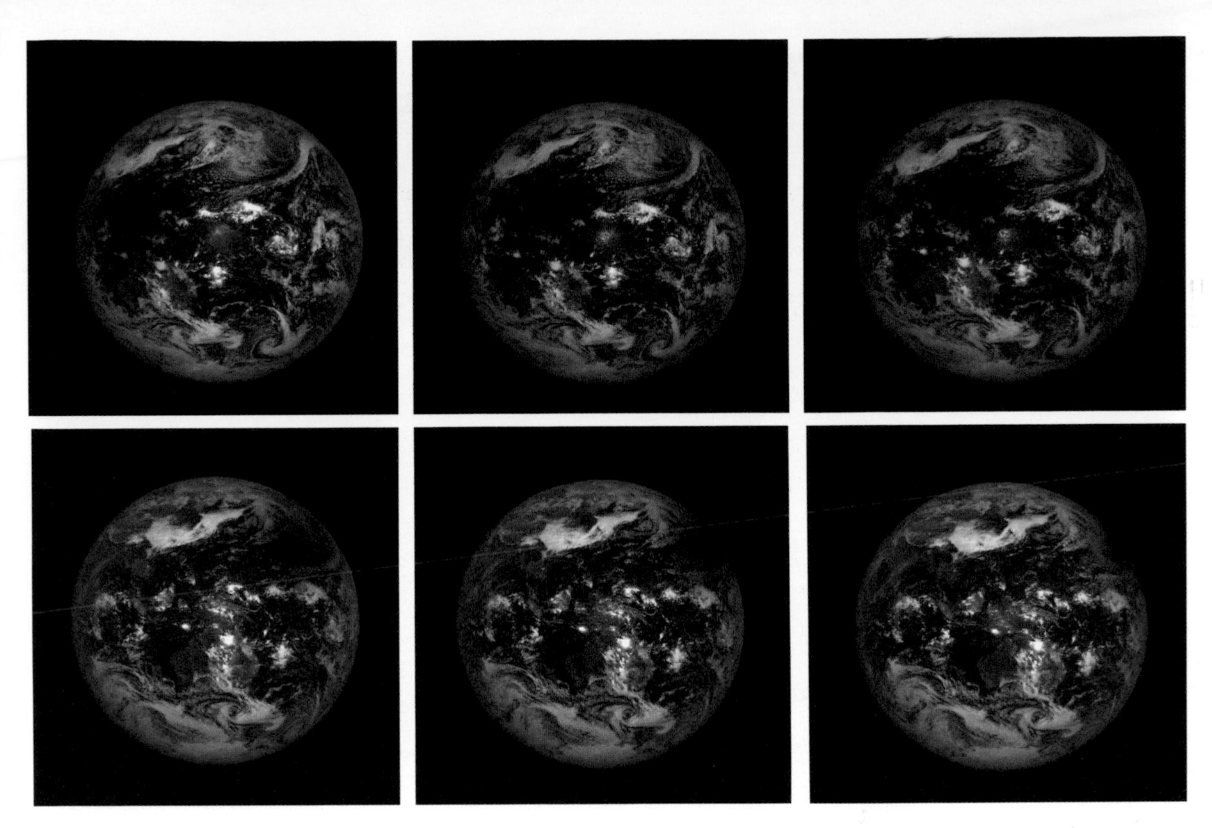

Image set credit: NASA EPIC Team

A Solar Eclipse Moving Across the South Pacific

Images above, as seen from the Deep Space Climate Observatory (DSCOVR) EPIC camera March, 2016. As the Moon passes between the Earth and the Sun, it casts a shadow, called an umbra, where a total eclipse is seen. Some regions will experience a partial eclipse, where the Sun's light is only partially blocked. This area is called a penumbra.

Image credits: NASA/Rami Daud

Super Blood Lunar Eclipse

A lunar eclipse occurs when the Sun, Earth, and Moon align so that light from the Sun is shadowed by the Earth. Lunar eclipses happen in the full-moon phases of the Moon's cycle. The Moon, whose phases take a month to cycle through, appears to take less than six hours, going from a visibly full Moon to a dark disc to back to full. A total lunar eclipse happens up to three times a year. The average occurrence of a lunar eclipse is one and a half times each year. A lunar eclipse is safe to view directly with your eyes, unlike a solar eclipse.

The Earth's Shadow

During a lunar eclipse, the Moon travels through two parts of the Earth's shadow. It goes through the penumbra, which is a partial shadow. Light is still getting through—the shadow is soft and incomplete. Then it enters the umbra, putting the Moon in a complete shadow. As the Moon continues, it moves out of the complete shadow into the penumbra, and then completely out of the Earth's shadows into complete sunlight once again.

Since the Moon's day is a month long, it sinks into and out of its cool dark night gradually taking two weeks. In contrast a lunar eclipse can quickly lower the temperature on the Moon's effected surface. People or instruments that work on the Moon need to have strategies in place for these abrupt changes.

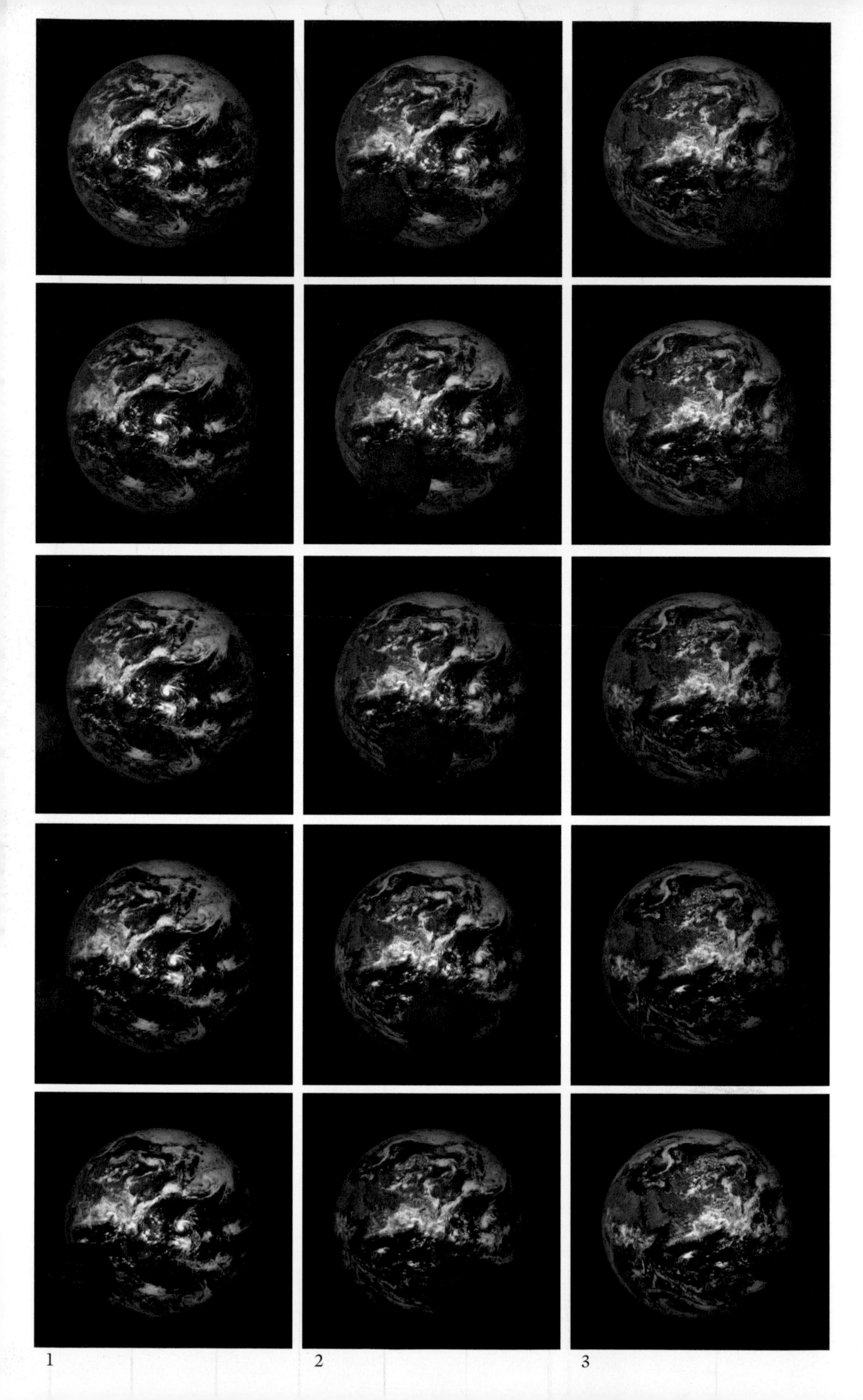
1
2
3

Image set credits: NASA/ DSCOVR

Lunar Transit

On July 5, 2016, the Deep Space Climate Observatory (DSCOVR) recorded the Moon's transit across the Earth as it passed. Although a lunar transit can't be viewed from Earth's surface, the DSCOVR makes seeing this truly magnificent event possible.

Image credit: NASA

Almost 4 Million Miles Away

NASA's Galileo spacecraft took this image of the Earth and Moon on December 16, 1992, from almost 4 million miles away (6.2 million km). Notice that both spheres are partially lit by the Sun and have the same half-illuminated appearance. If the craft was between the Earth and the Sun, the spheres would have been fully illuminated.

A View from Almost 40 Million Miles Away

NASA's OSIRIS-REx spacecraft captured the Earth and Moon *(facing page)* from almost 40 million miles away on January 17, 2018. The pair are in the middle of the image. The larger bright spot on the left is Earth. The smaller spot is the Moon.

Several constellations are faintly visible in this image. The Pleiades, in the Taurus constellation, can be seen in the upper left corner. In the upper right corner *(facing page)*, the brightest star in Aries, Hamal, can be seen. The Earth and Moon are placed within five stars that are the head of Cetus the Whale, who in Greek mythology is a sea monster.

Exomoons

Exoplanets, planets that exist outside of our solar system, are detected using transit photometry and Doppler spectroscopy. So far, no exomoons have been detected on these planets.

Image credits: NASA/Goddard/University of Arizona/Lockheed Martin

Early Lunar Observations & Humanity's Narratives

Human Prehistory and the Moon

The Moon has existed for over 4 billion years, having formed not long after the formation of Earth. Needless to say, it has been in the sky before humans or any life forms existed on the planet. Early humans may not have known what the Moon was made of, but they certainly were very familiar with its presence and celestial conduct.

The Moon as a Timepiece

The regularity of the Moon's movement across the daily sky, monthly phase changes, and seasonal arc changes have provided a valuable and dependable timepiece throughout the ages. Along with the Sun, the Moon has helped to mark the coming and going of the days. The seasons were delineated into monthly segments convenient for communicating the passing of time. For example, "Let's meet again in one moon" or "The baby is due soon—it's been nine moons."

Just as important, when people understood and used the Moon's cycles to comprehend the seasons, the gathering and growing of food became a predictable and successful endeavor.

Even with evermore sophisticated chronometers *(facing page)*, the Moon is included even if only for aesthetic enhancement. The Moon is a part of humanity's time keeping.

Image credit *(facing page and below)*: public domain

The Man on the Moon

We can assess "the Man in the Moon" as a pareidolic phenomenon. When looking at the Moon's disc, one sees shapes that are vague and suggestive. They evoke an interpretation by the observer of something they are familiar with, such as a face. This is enough to initiate legends, stories, rhymes, and riddles.

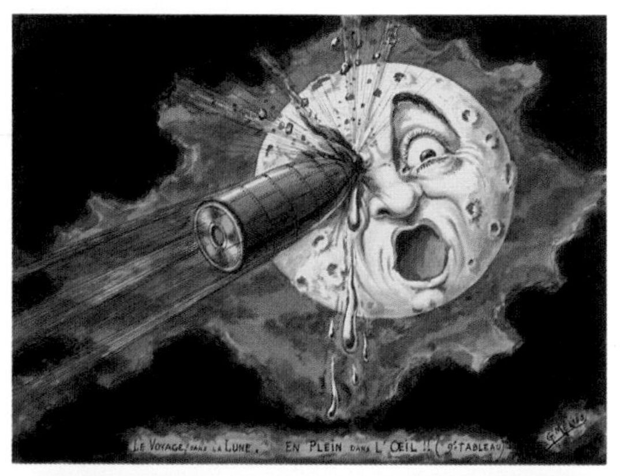

The Moon and Science Fiction

It's only natural to speculate about the Moon. What is it made of? How did it get there? Of course, it's in the sky, but really, where is it? And, does anyone live there? The first stories were probably creation myths. As humanity's observations, knowledge, and technology progressed, so did fiction's sophistication.

Image set credit: public domain

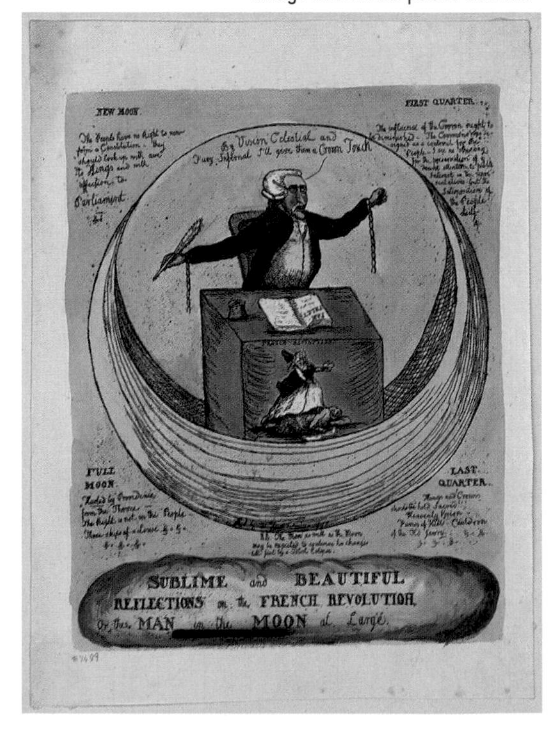

The First Men on the Moon

These illustrations, by Henri de Montaut were made for Jules Verne's 1865 novel *From the Earth to the Moon*. The book is speculation about traveling to the Moon: how to accomplish it, rudimentary calculations, what would be found, and the possibility of intelligent life.

Early Lunar Observations

Prehistoric records of the Moon's cycles have been found on bone and antler, the earliest dated to 32,000 BC.

Babylonian astronomer's systematically recorded their celestial observations *(top right)*, possibly as early as about 700 BC. The cycle of lunar eclipses was observed, and at one point they were able to create mathematical models that would predict the eclipses.

The invention of Galileo's telescope revolutionized observations of the Moon and the stars. Galileo Galilei published his astronomical treatise *(bottom right)* in 1610. He made observations of the Moon *(facing page)* that showed mountains, valleys, craters, and shadowing of its sphere, indicating its phases.

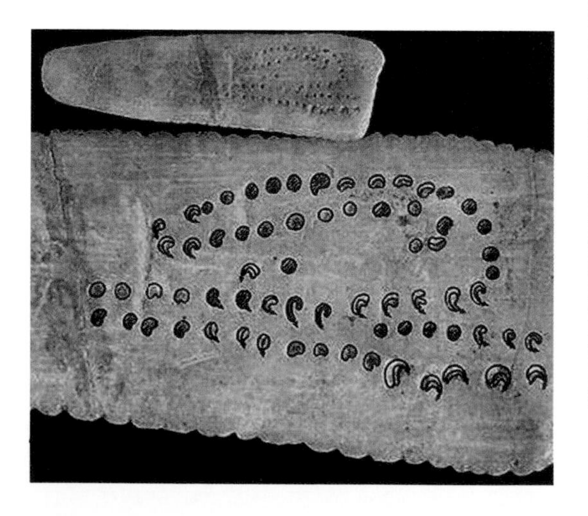

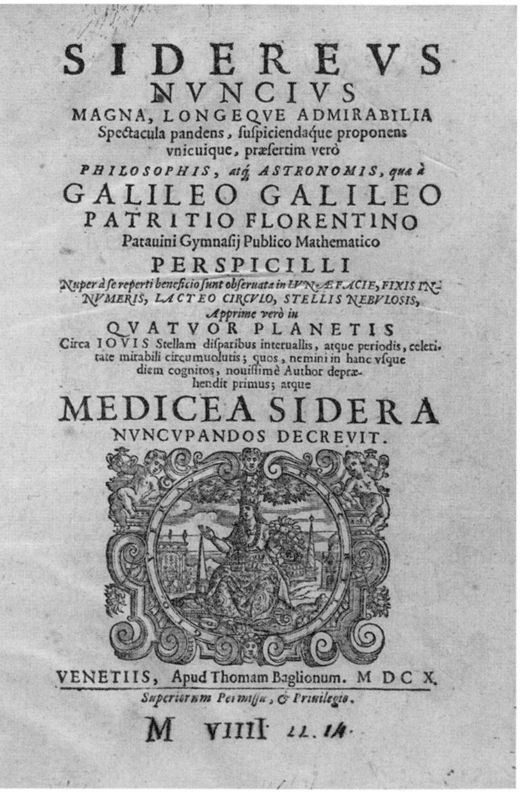

SIDEREVS
NVNCIVS
MAGNA, LONGEQVE ADMIRABILIA
Spectacula pandens, ſuſpiciendaque proponens
vnicuique, præſertim verò
PHILOSOPHIS, atq; ASTRONOMIS, quæ à
GALILEO GALILEO
PATRITIO FLORENTINO
Patauini Gymnaſij Publico Mathematico
PERSPICILLI
Nuper à ſe reperti beneficio ſunt obſeruata in LVNÆ FACIE, FIXIS INNVMERIS, LACTEO CIRCVLO, STELLIS NEBVLOSIS,
Apprime verò in
QVATVOR PLANETIS
Circa IOVIS Stellam diſparibus interuallis, atque periodis, celeritate mirabili circumuolutis; quos, nemini in hanc vſque diem cognitos, nouiſſimè Author depræhendit primus; atque
MEDICEA SIDERA
NVNCVPANDOS DECREVIT.

VENETIIS, Apud Thomam Baglionum. M DC X.
Superiorum Permiſſu, & Priuilegio.
M VIIII

Image set credit: public domain

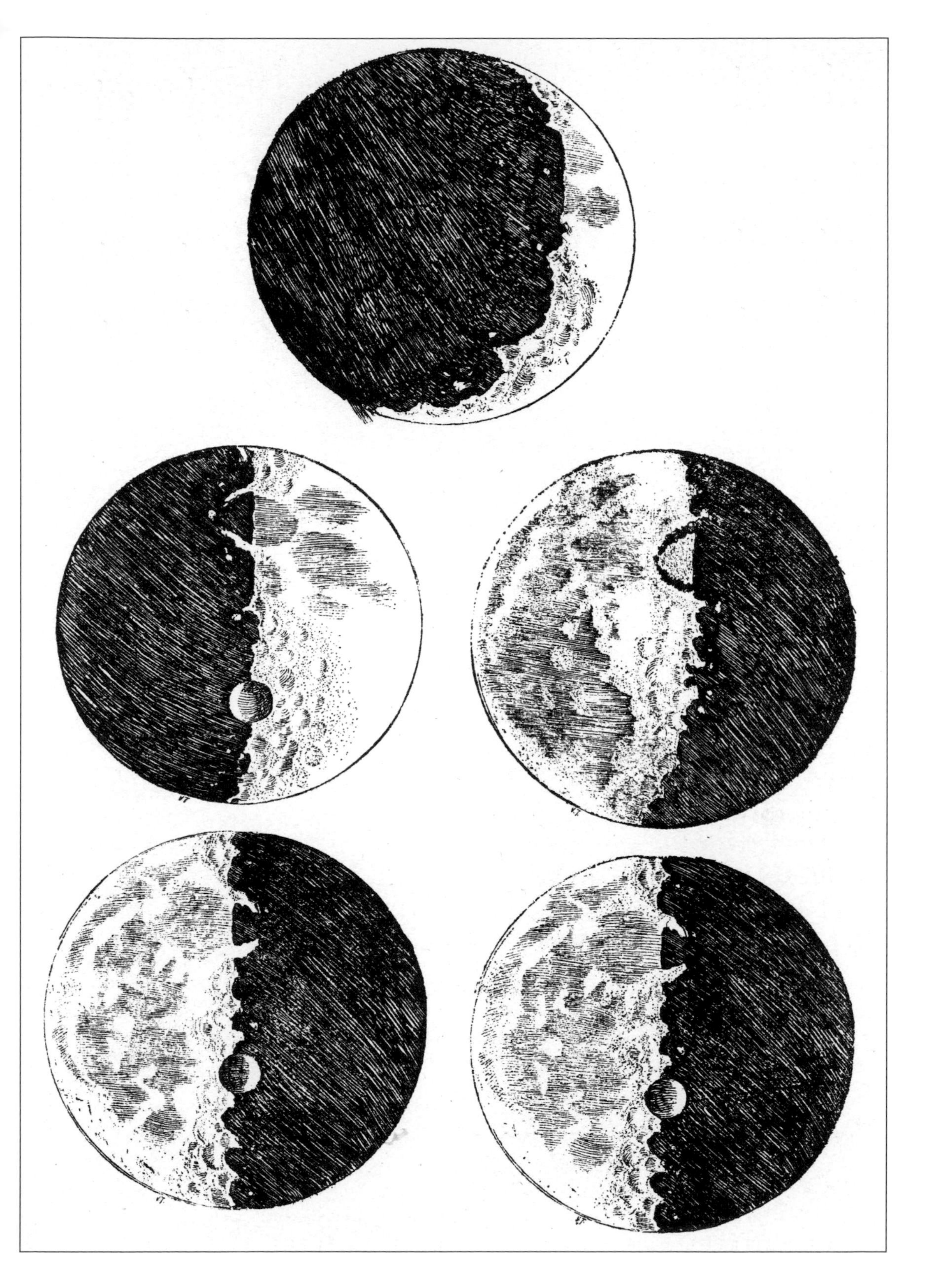

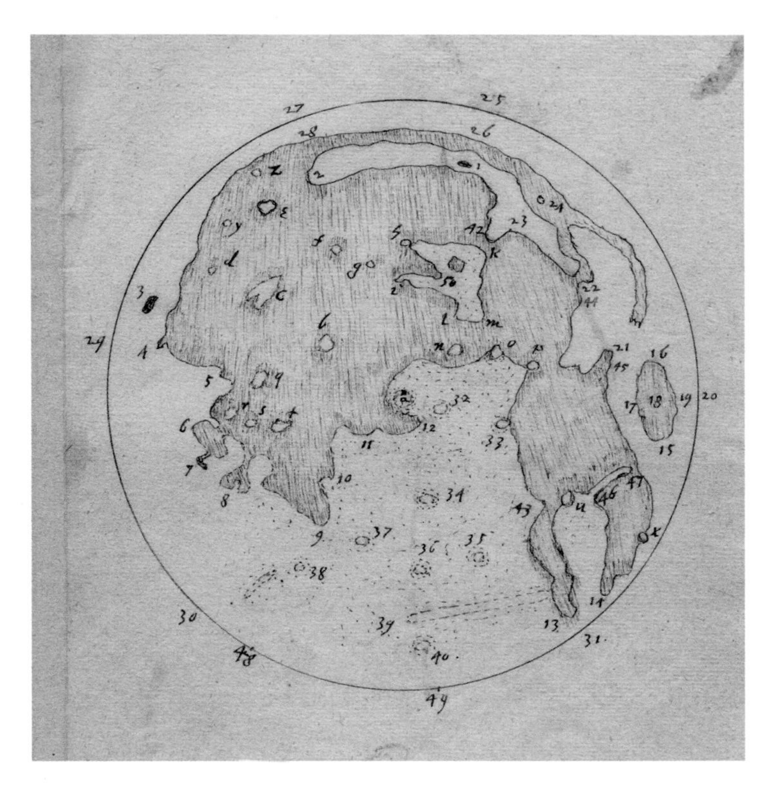

First Recorded Observations with a Telescope

Thomas Harriot was a contemporary of Galileo. Galileo's descriptions were published first and are thought to be more

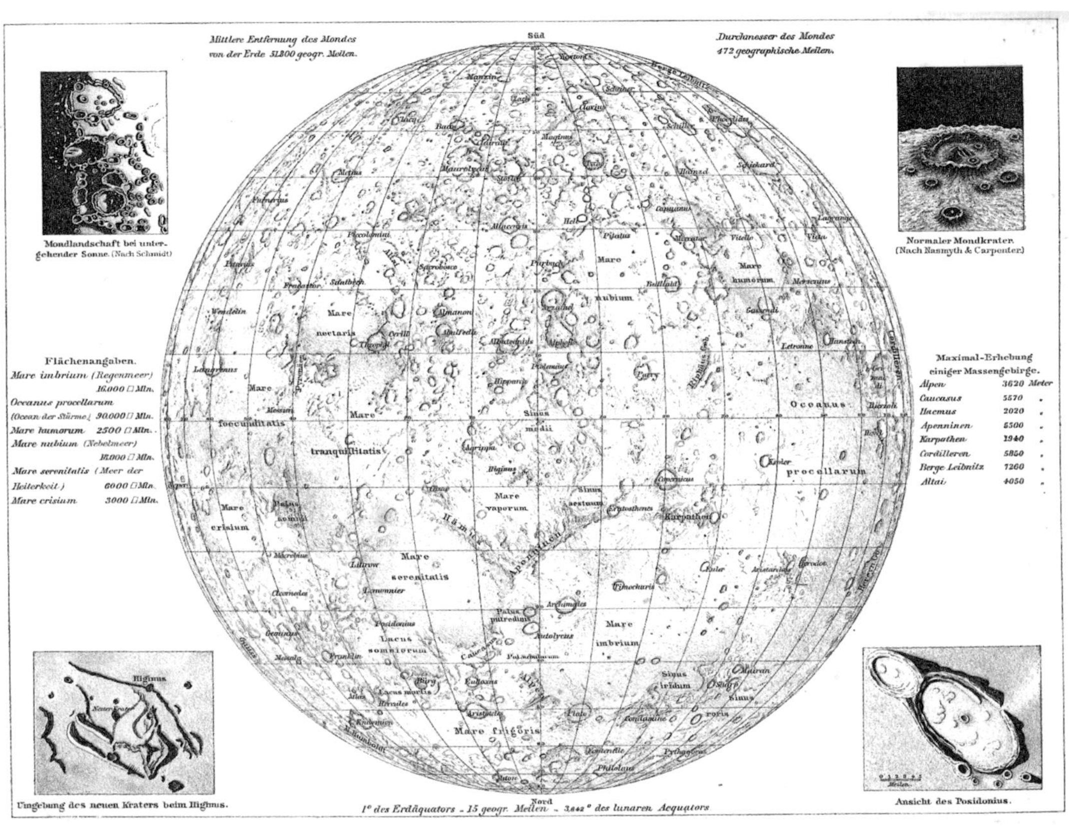

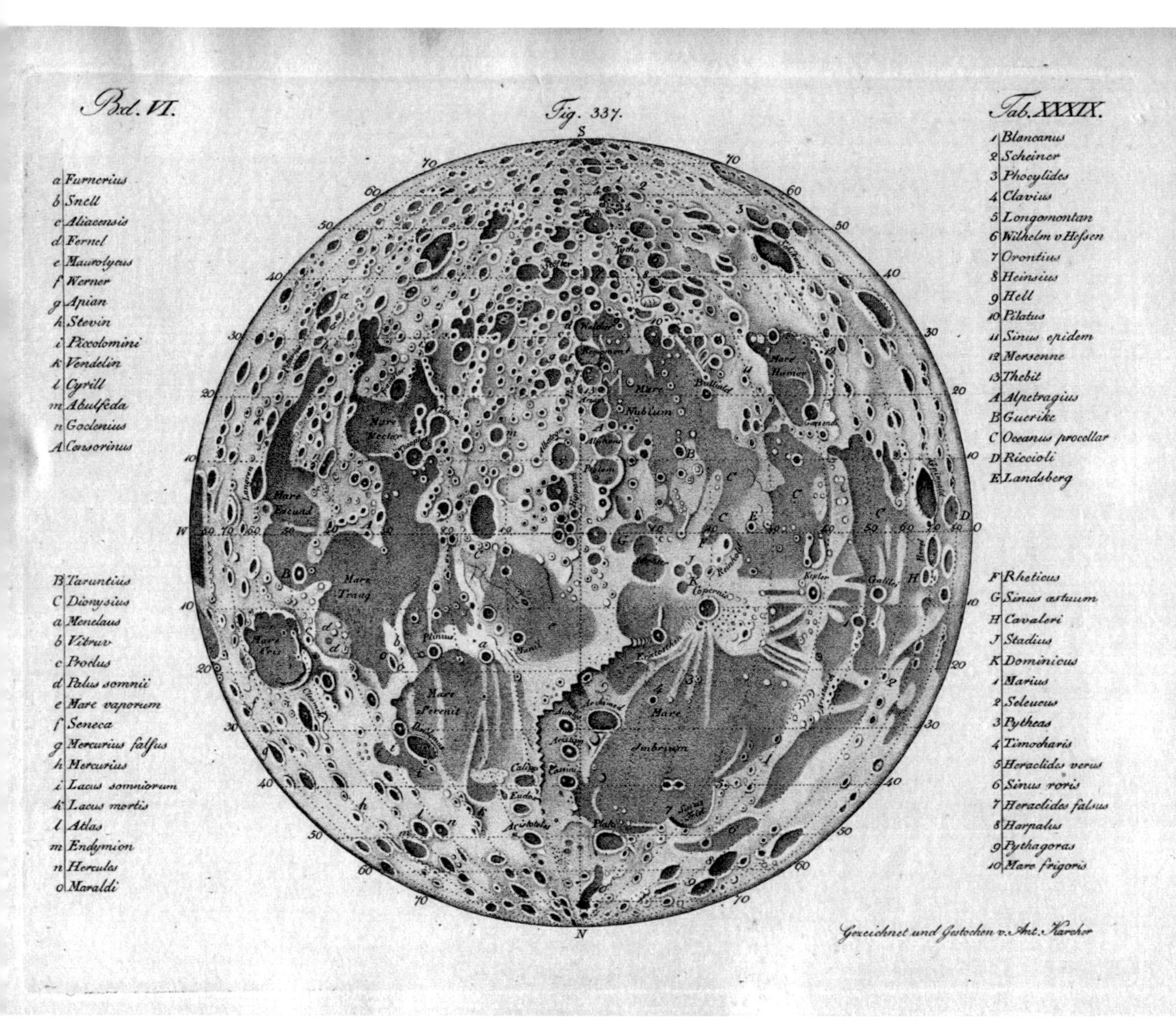

topographically descriptive. In spite of not publishing first, Harriot is now recognized as the first person to make illustrations of the Moon *(facing page, top)*, based on his observations with a telescope. Harriot's descriptions incorporate useful cartographic concepts used in charting and mapping spheres.

As telescopes improved, so did the maps of the Moon. Compare this Moon map *(top)*, made in 1840, with this detailed map of the Moon's near side *(facing page, bottom)*, by Richard Andree, was published in 1881.

Image credits *(facing page, bottom)*: Map of the Moon from the Andrees Allgemeiner Handatlas (1881) by Richard Andree

The Apollo Program

Apollo Missions

The United States successfully landed the first humans on the Moon between 1969 through 1972. There were six spaceflights with twenty-four men who left Earth and orbited the Moon. Twelve of these men walked on the Moon.

The Lunar Orbit Rendezvous (LOR) used the Apollo command and service module *(bottom*

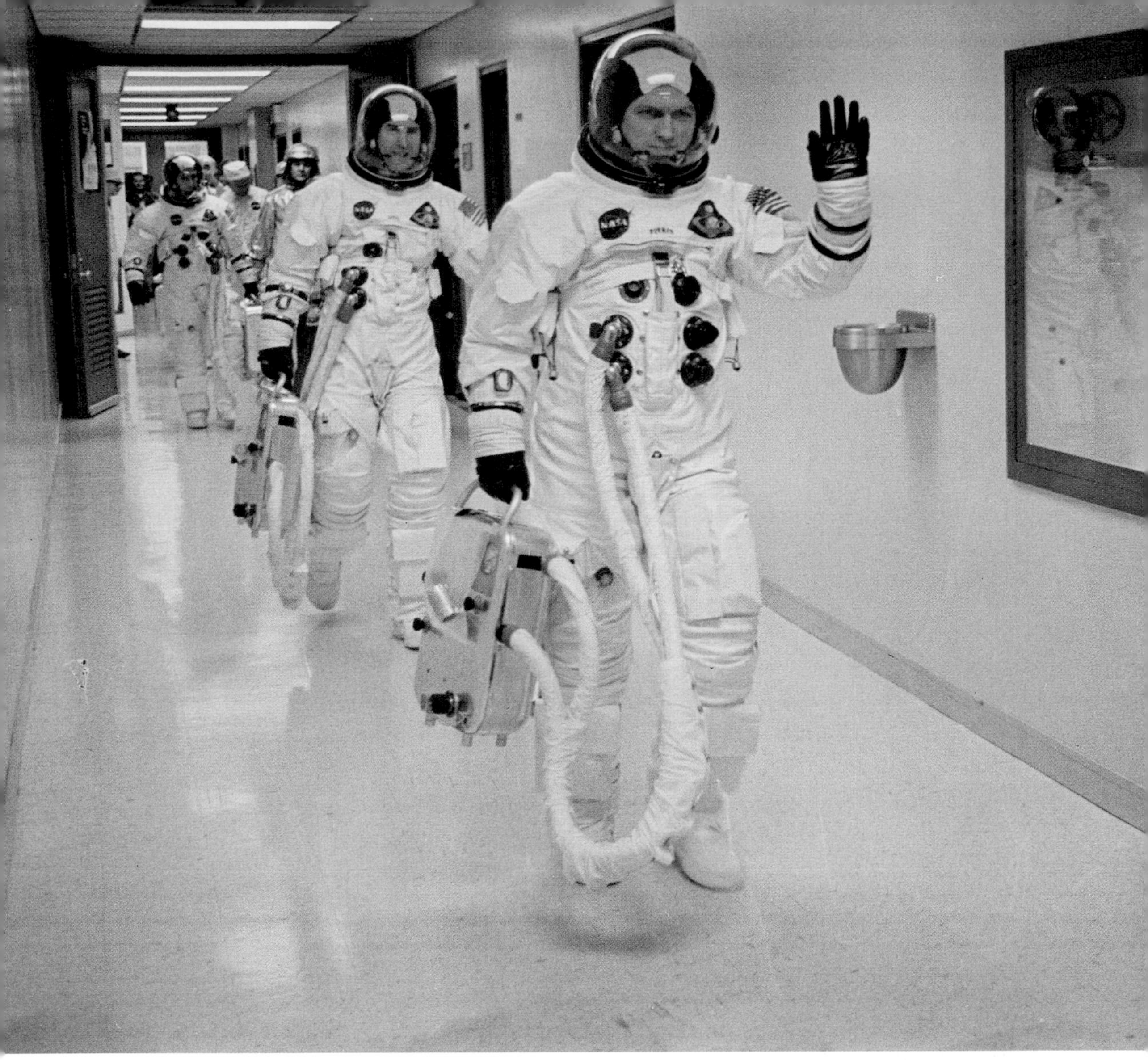

Image set credit: NASA

right), which stayed in the Moon's orbit, while a two-stage lunar module descended to the surface. Apollo 17 was the last crewed mission to the lunar surface. This crew spent the longest time on the Moon at just over three days. All of these missions that landed on the Moon successfully returned *(bottom right)*.

Image credit *(above and facing page)*: NASA

Traveling to the Moon

Apollo first started their lunar endeavors in the 1960s through the 1970s, when technology was much less advanced than it is today.The challenges are the same, but our tools are more sophisticated and refined. The world is technologically prepared for tasks of upcoming missions.

Walking on the Moon

The field of robotics has advanced since men have walked on the Moon. Today we send machines in before human presence is necessary. However, the fact that humans have visited and walked on the Moon is awe inspiring. If nothing else was accomplished, it raised our expectations, made us aware of human abilities, and urged us forward. Every one of us knows humans will walk on the Moon again.

Image set credit: NASA

Lunar Science

The Apollo experiences and the data collected are still being studied today.

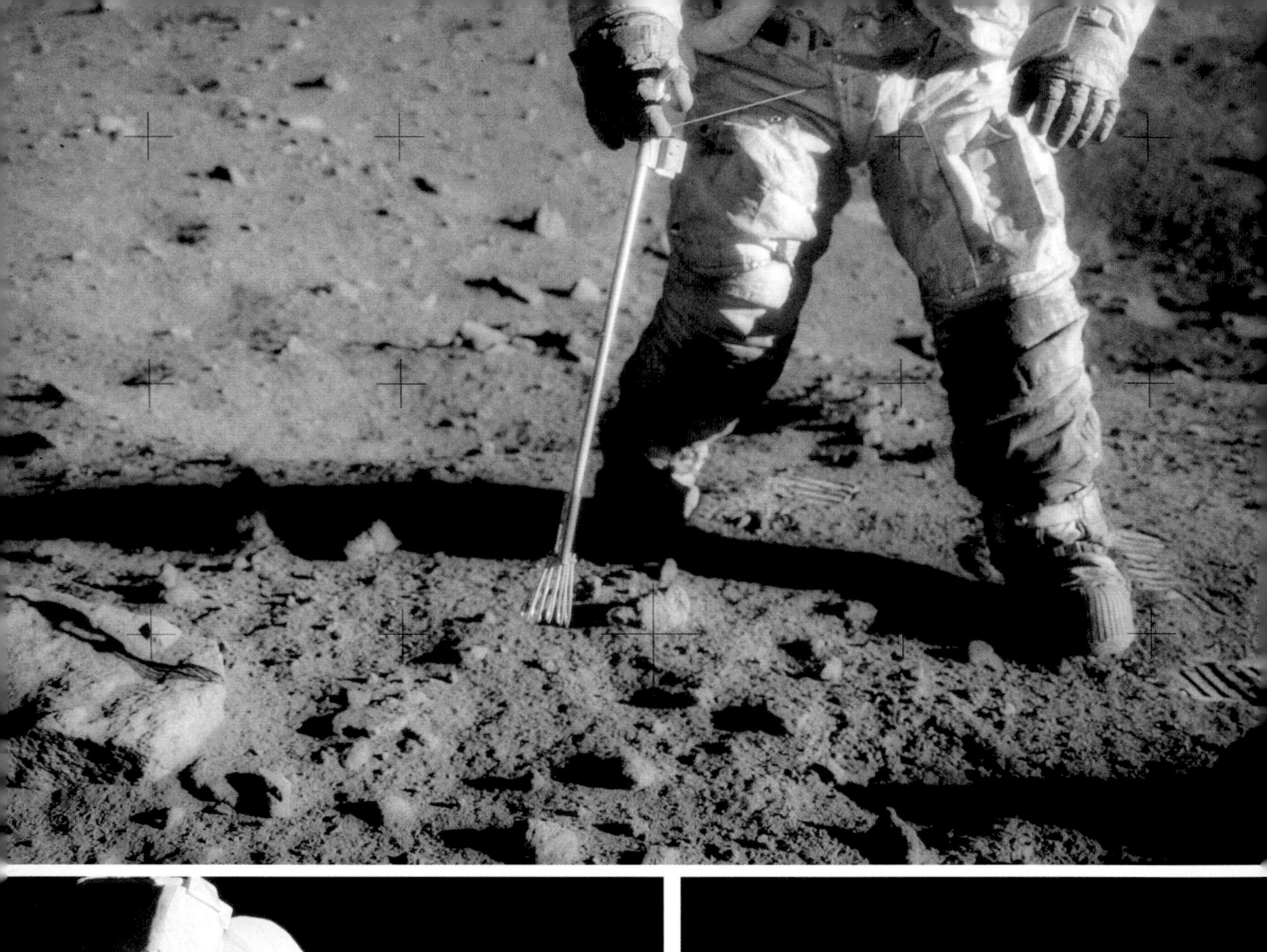

Image set credit: NASA

A Giant Leap

Neil Armstrong spoke of “That’s one small step for a man, a giant leap for mankind” when he first stepped on the Moon in 1969. The Apollo mission astronauts’ footprints are on the lunar surface still, waiting for our return as we prepare for our first steps to Mars.

International Exploration

Space Law

Space law is challenging because the boundaries in space are difficult to define. These laws will need to be addressed when questions arise about legal jurisdiction of spacecraft orbiting Earth and other celestial bodies. The laws will also reflect international agreements on policies and treaties regarding space exploration. The United Nations General Assembly created the Committee on the Peaceful Uses of Outer Space (COPUOUS). The committee currently has 92 members. They promote the peaceful use of space, cooperation, and the sharing of information for exploration.

Space law is important enough that now, several universities have programs in it and give degrees. Space law deals with many subjects, including: property rights, information sharing, technology sharing, weapons and military use, commercial use, orbit allocation, protection of astronauts, liability for damage, and environmental preservation. The law is further entwined in other kinds of law that deals with administrative law, intellectual property, insurance, criminal and commercial law, and so on. Needless to say, space law is, and will continue to be challenging. Treaties, and international and domestic agreements, rules, and principles will all be necessary and be constantly revisited to make the human space enterprise possible. These treaties cover rescue agreements, registration of objects launched, ethics, and more, for the Moon and other celestial bodies.

International Explorations

The endeavor of exploring space is enormous. International ventures in exploration bring together skills, resources, and technological circumstances from a world-wide base. The nations and entities that have reached the Moon are the former Soviet Union, the United States (NASA), the European Space Agency (ESA), Japan Aerospace Exploration Agency (JAXA), and China National Space Administration (CNSA).

The first images of the far side of the Moon *(page 58)* were made by the Soviet Union's Luna 3 craft in 1959. The Luna program was a series of robotic missions sent to the Moon between 1959 and 1976. Luna 24 in 1976 returned from the surface with lunar soil samples.

The United States NASA Apollo missions *(pages 44-51)* were carried out from 1961 to 1972 with the first crewed flight in 1968. During these missions, 24 men flew to the Moon and 12 men landed on the lunar surface during six landings. Lunar rock and soil were returned to Earth for study.

The Institute of Space and Astronautical Science of Japan launched the Hiten *(right top)* spacecraft in 1990 and stayed in orbit until a deliberate crash in 1993. In 2007, Selene, the second Japanese lunar orbiter spacecraft was launched.

The European Space Agency's SMART-1 *(upper middle)* a Swedish-designed satellite was launched in 2003. Its orbital maneuvers used ion engines, employing electricity generated by solar panels.

Chang'e 1 launched in 2007, was China's first craft in a series of lunar explorations *(see pages 54-55)*.

The Indian Space Research Organization (ISRO) sent the Chandrayaan-1's Moon Impact Probe in 2008. As a result, water was discovered. The Chandrayaan-2 is an orbiter, lander, and rover developed by the ISRO with a 2019 planned mission date.

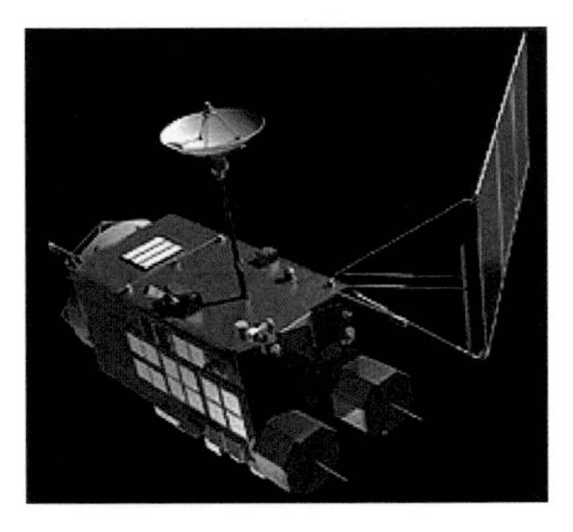

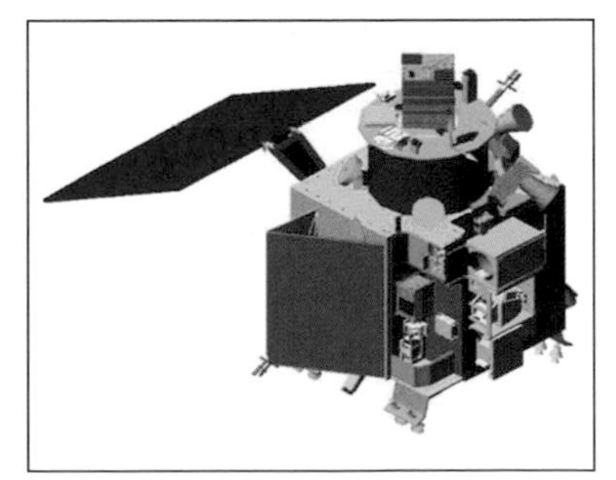

Image set credit: NASA

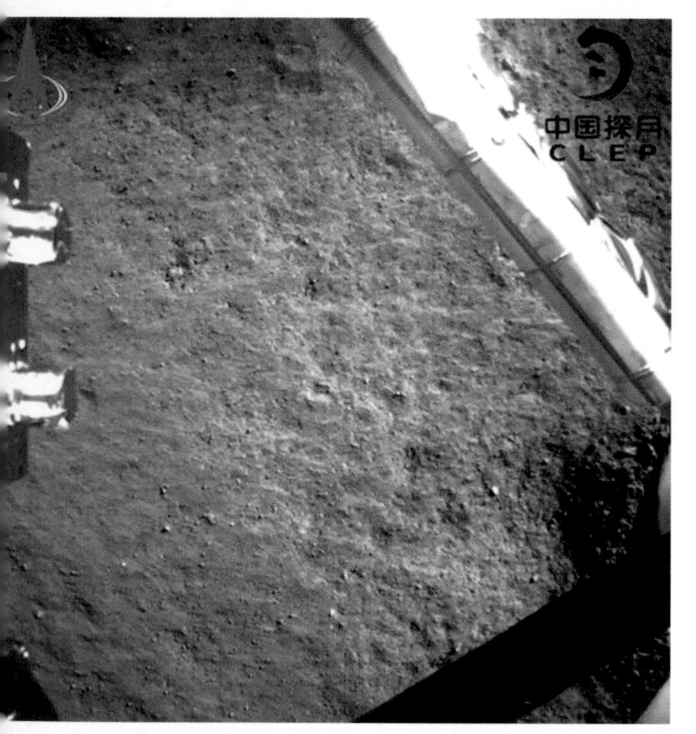

Chang'e Lunar Landing Site

The China National Space Administration's (CNSA) Chinese Lunar Exploration Program (CLEP) successfully landed Chang'e 4 on the far side of the Moon *(facing page)* on January 3, 2019. The landing area *(facing page bottom left and right)* is near Von Kármán, a large lunar impact crater that is located in the southern hemisphere, which is located within an even larger impact crater known as the South Pole–Aitken basin. This larger crater is one of the largest impact craters in the solar system.

This is China's second mission to land and rove on the Moon. These are some of the first images *(middle and bottom left)*. There is a robotic lander *(facing page middle left)* and a rover *(facing page middle and top right)* called Yutu 2, which means "Jade Rabbit No. 2." The lander goes into temporary hibernation when the landing area goes into its lunar night, becomes dark, and loses its solar energy.

The tasks for this mission are to determine the age and composition of this unexplored region of the Moon. China has international partners, including Sweden, Germany, the Netherlands, and Saudi Arabia, that have partially supplied payloads.

Plans for the future include Chang'e 5 and Chang'e 6 missions to collect samples and send them back to Earth. As with other nations, more missions beyond these are also planned.

Image credit *(top, middle left and right, and facing page image set):* China National Space Administration (CNSA)

Image credit *(bottom left and right):* NASA

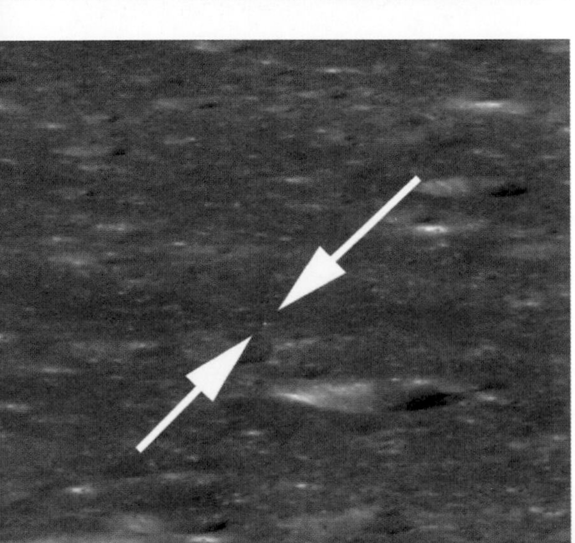

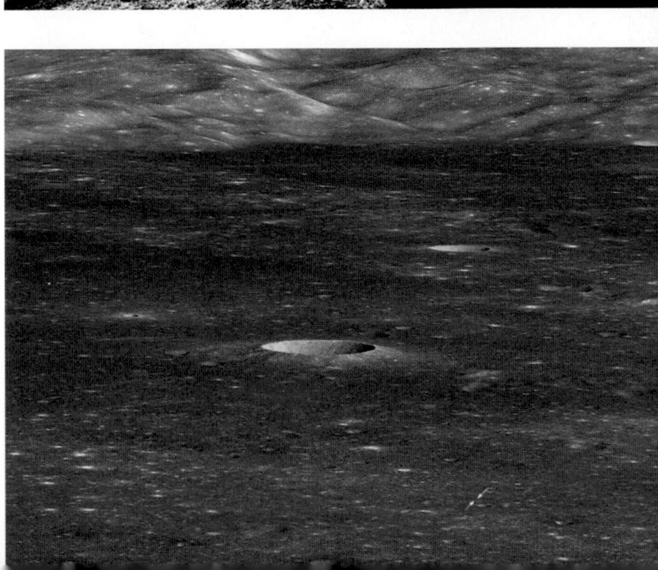

Lunar Geography

Exploring the Moon from Afar

The geography of the Moon is the study of the surface, features, inhabitants, and phenomena. Short of a crew arriving in a spacecraft, we can study the features of the Moon's facing surface from the Earth. Study of the far side requires remote satellites and probes.

Many things can be discovered about the Moon from a distance. For example, the reflectivity or albedo of an area of the surface indicates the likelihood of mineral make up, such as titanium or another mineral makeup with more iron. Knowing the albedo helps the lunar geologists differentiate between areas that have received space weathering—caused by solar wind and tiny meteorite impacts—from other features and geologic processes.

One area of geography is focused on mapping the features on the surface, naming them, and discovering related processes and phenomena.

Image *(right):* Lunar Day from *Recreations in Astronomy* by Henry White Warren, D.D. 1879

Image credit *(facing page):* Giuseppe Donatiello Moon albedo at IR 1000nm (germanium filter)

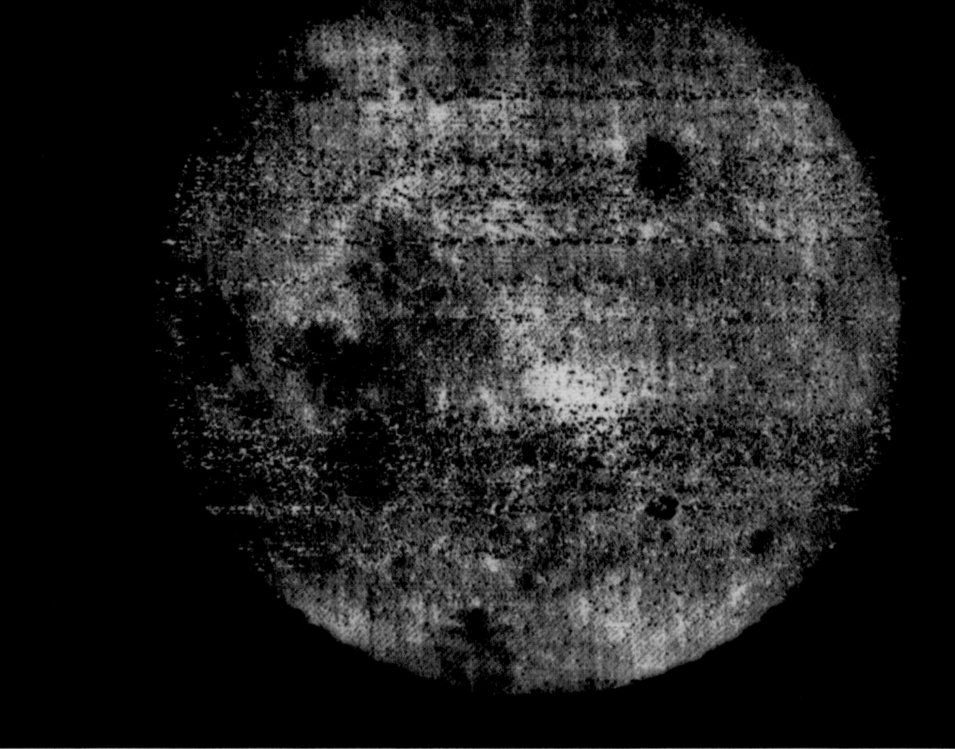

First View from the Far Side of the Moon

We don't see the far side of the Moon from Earth because of its synchronous rotation. The image *(above)* taken in 1959 by Luna 3 is one of the first images taken of the far side of the Moon. Luna 3 was the third successful Lunar spacecraft returning with the first ever images of the far side.

Image credits: NASA, J. Bell (Cornell U.) and M. Wolff (SSI)

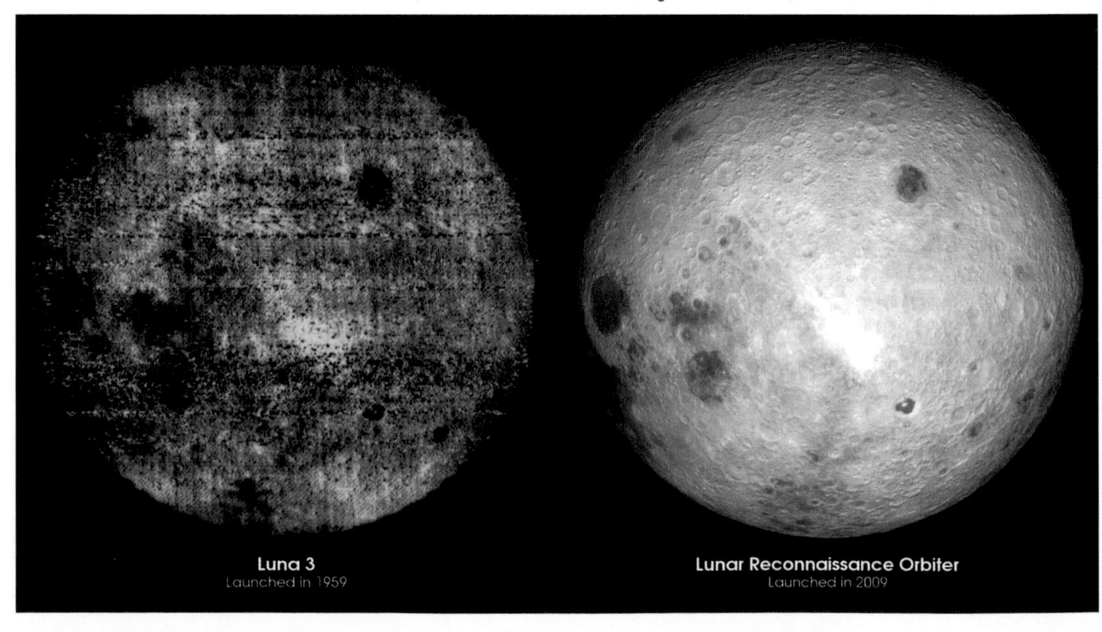

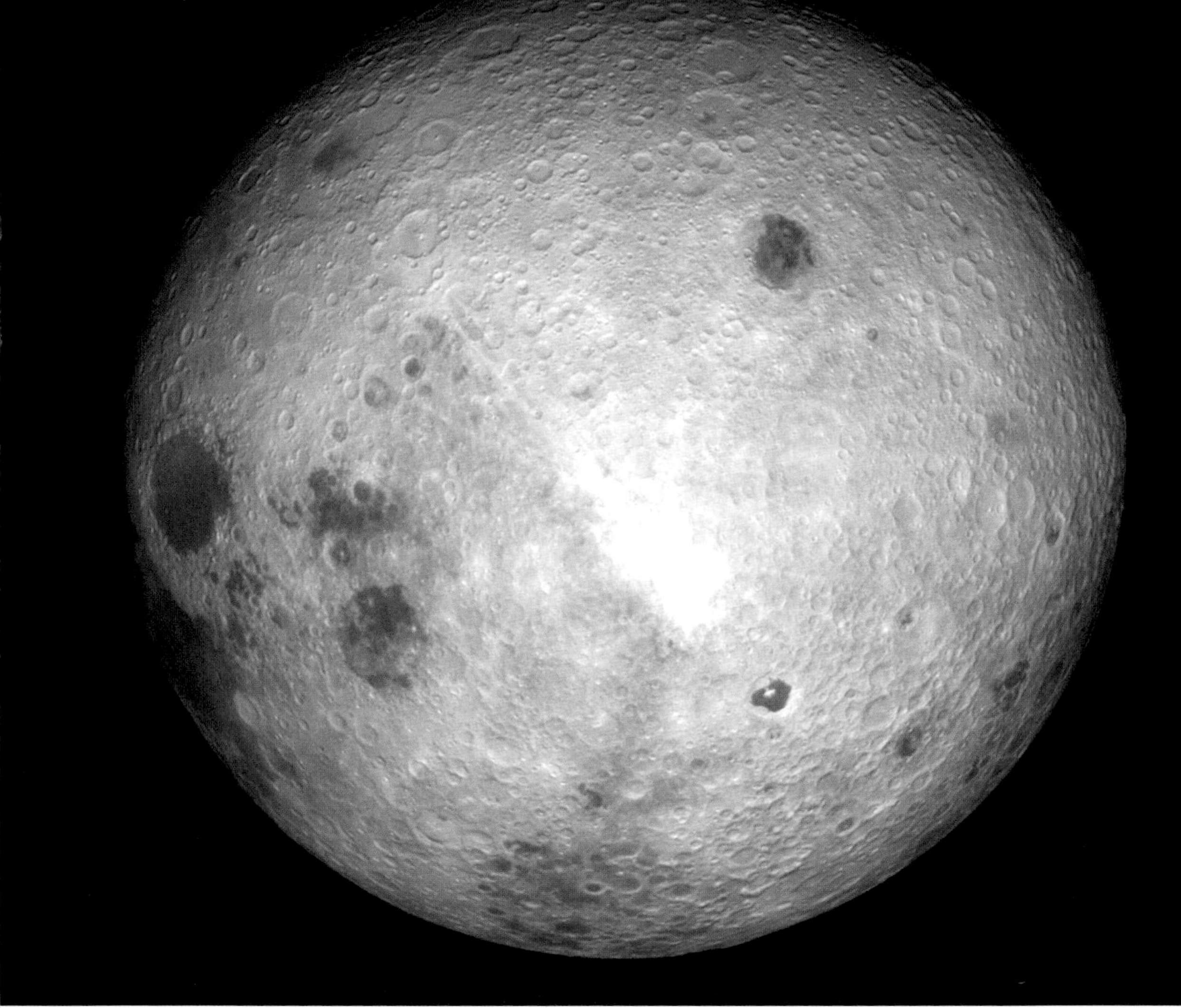

Image credit: NASA

Remote Viewing

Today, we have very detailed images *(above)* of the Moon's far side from the Lunar Reconnaissance Orbiter (LRO). Compared to the first acquired indistinct images from Luna 3*(facing page, bottom)*, these new images show the features with much greater detail.

Apollo astronauts, Frank Borman, William A. Anders, and James A. Lovell Jr., were the first humans to witness the far side of the Moon during their Apollo 8 flight.

Image credit: National Space Science Data Center

Landing Sites

A Moon landing site is where any spacecraft has reached the surface of the Moon. Luna 2 of the Soviet Union's space program was the first craft to arrive on September 13, 1959.

The landing sites in this image include the Soviet Union's Luna missions, and the United States' Surveyor and Apollo missions. Since these missions to the Moon, additional crafts have been sent by these two countries as well as other countries: Japan, European Space Agency, China, and India.

Apollo Landing Sites

The United States' Apollo missions were the first and only crewed missions to the Moon. Apollo 11 of the United States was the first to carry anyone in a spacecraft to land on the Moon on July 20, 1969. There were six landings in total: Apollo 11, 12, 14, 15, 16, and 17. The illustration above shows the six Apollo craft landing sites.

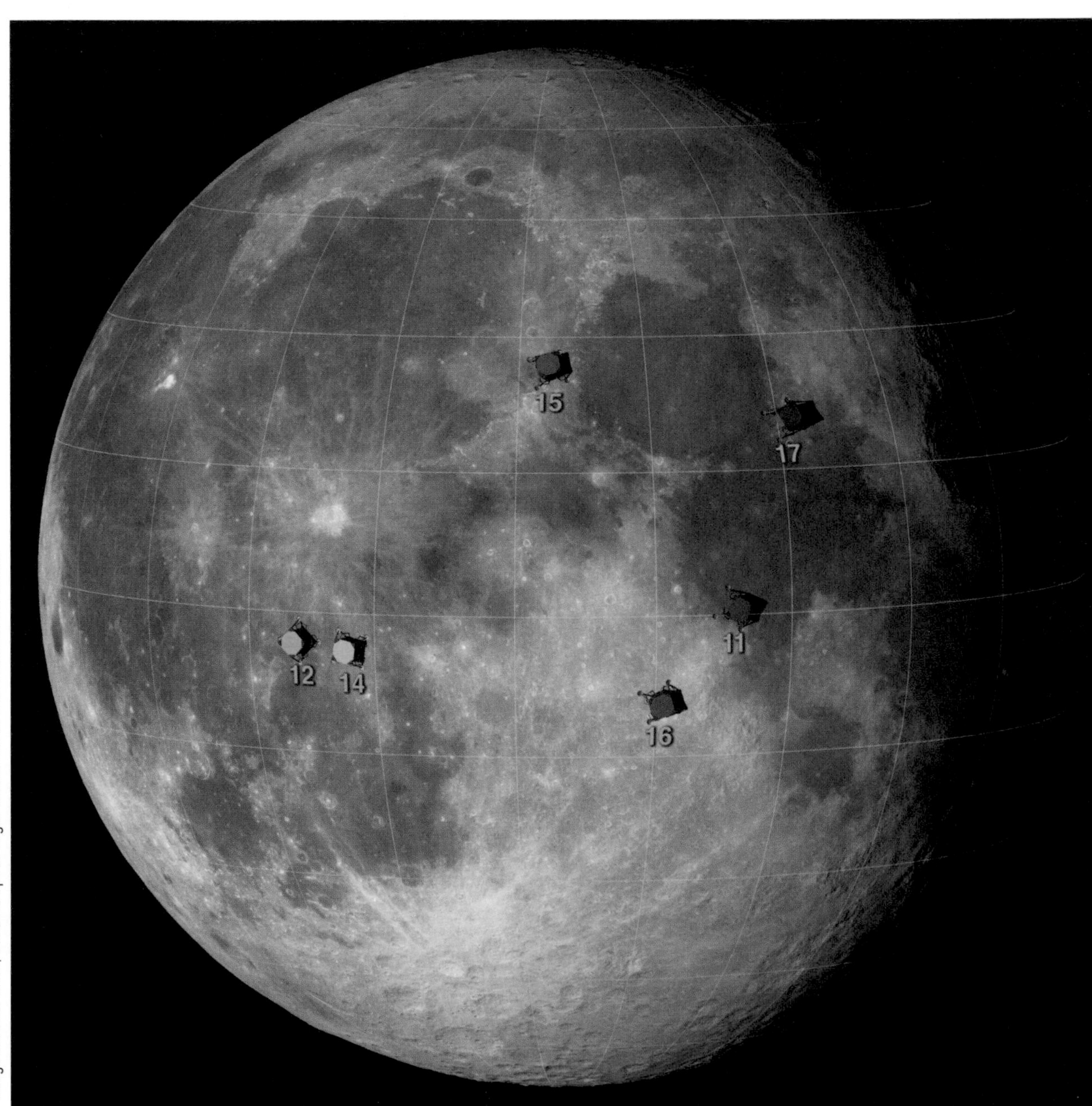

Image credits: NASA/Goddard Space Flight Center Scientific Visualization Studio

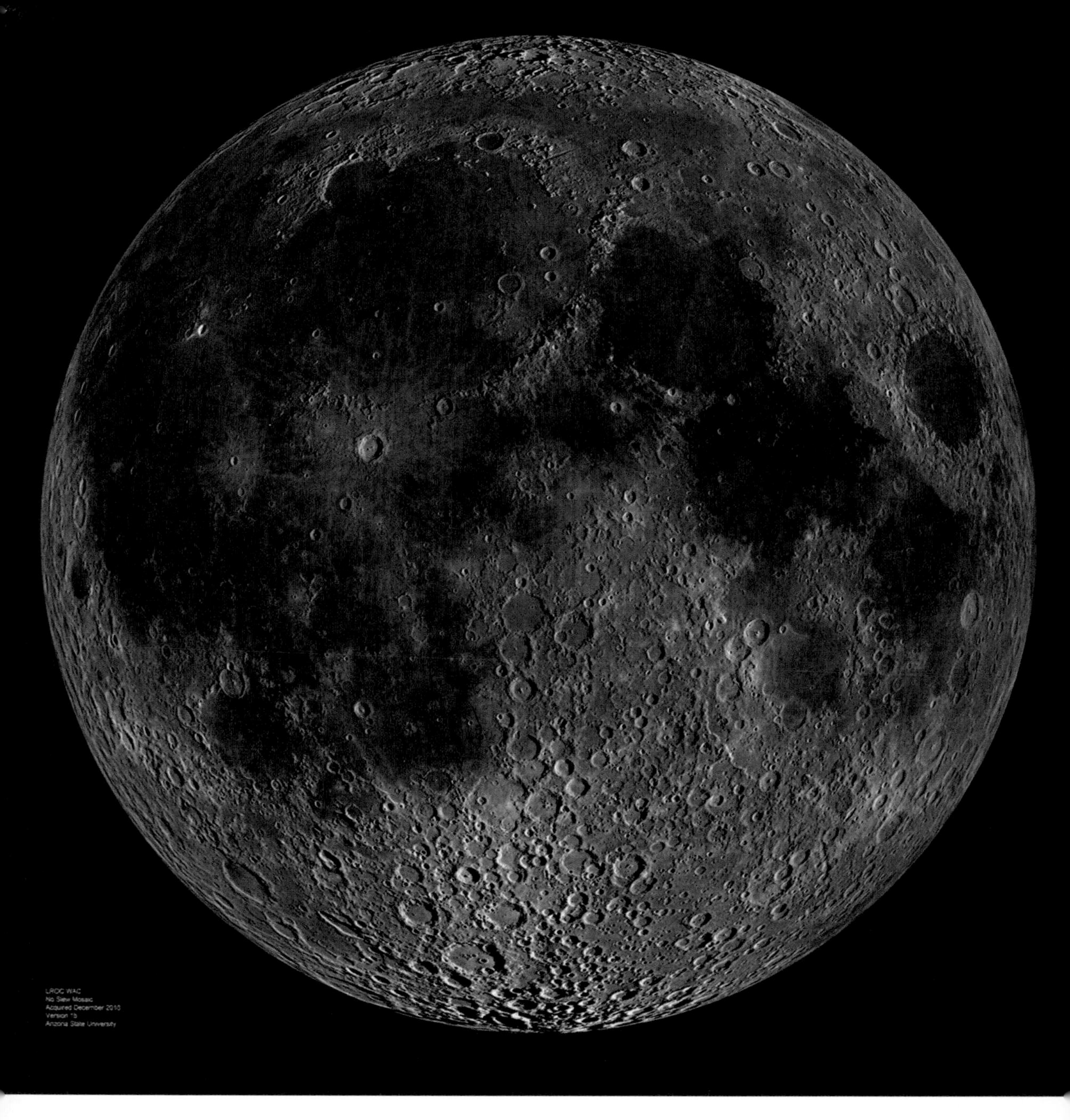

Near Side of the Moon

The near side of the Moon is sometimes referred to as the front side. It is tidally locked to the Earth—this side is always visible from the surface of the Earth. One theory about why the Moon is in this synchronous orbit is that the two sides are so different that the gravitational pull between Earth and the Moon's near side is stronger, locking it into synchronicity.

Far Side of the Moon

The far side of the Moon is sometimes referred to as the back side. This side was never seen until the Soviet Luna 3 space probe took pictures of it in 1959. This image of the far side was taken from the Apollo 16 spacecraft in 1972. Until images were obtained of the far side, most scientists had presumed this side would be much like the near side.

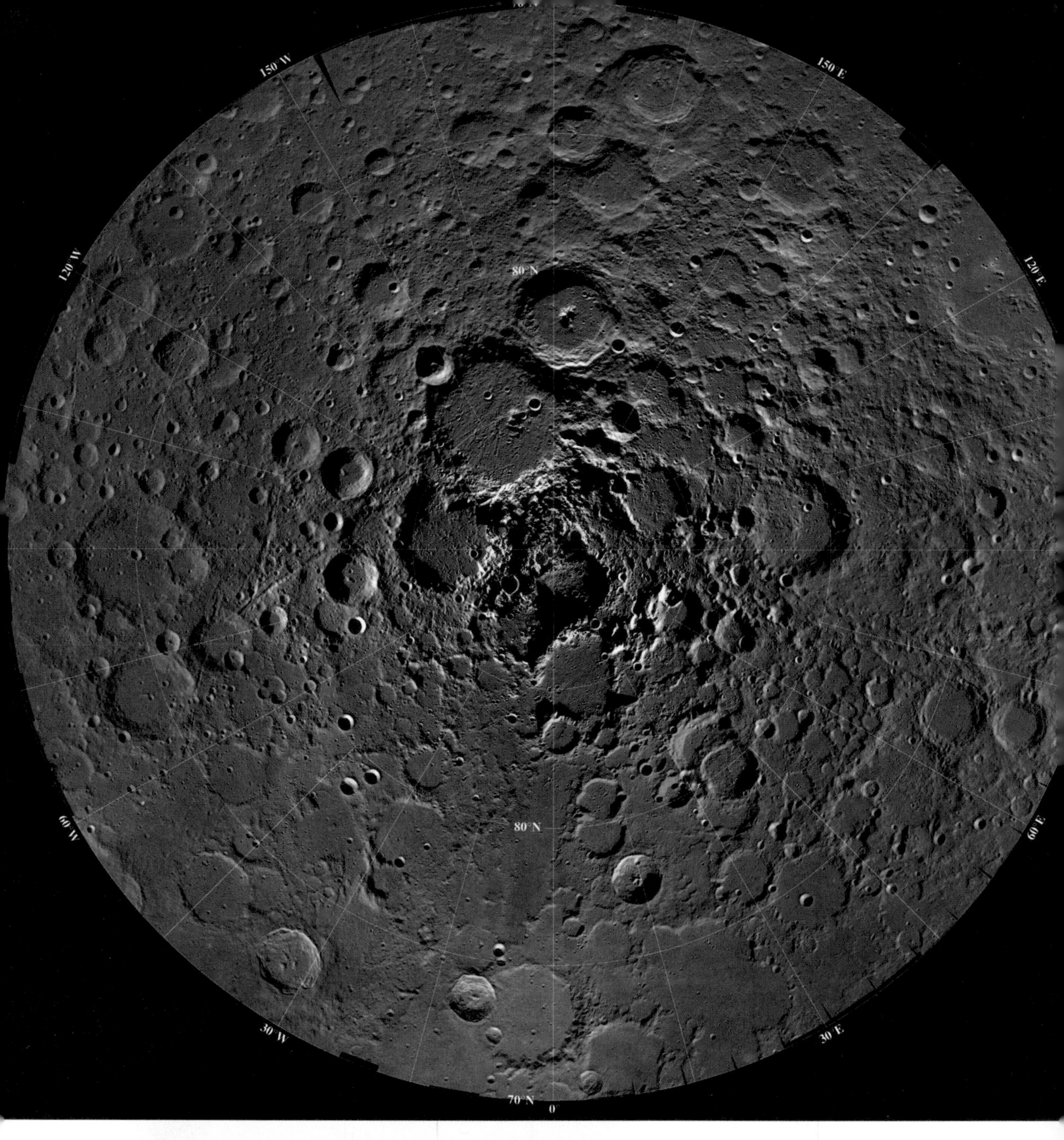

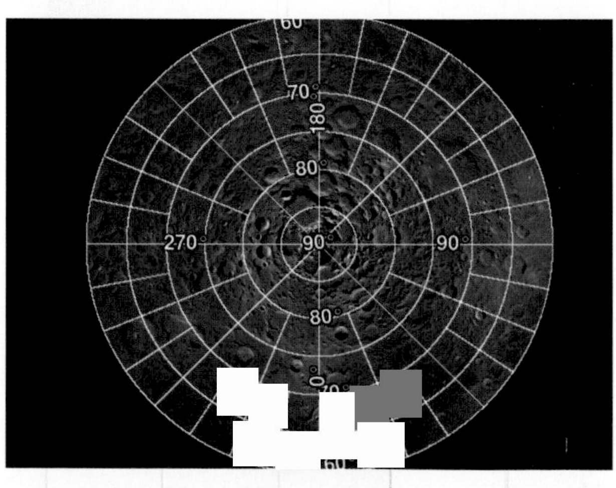

Lunar Polar Regions

The poles are determined by where the two points of a sphere's axis of rotation meet

Image credit: NASA

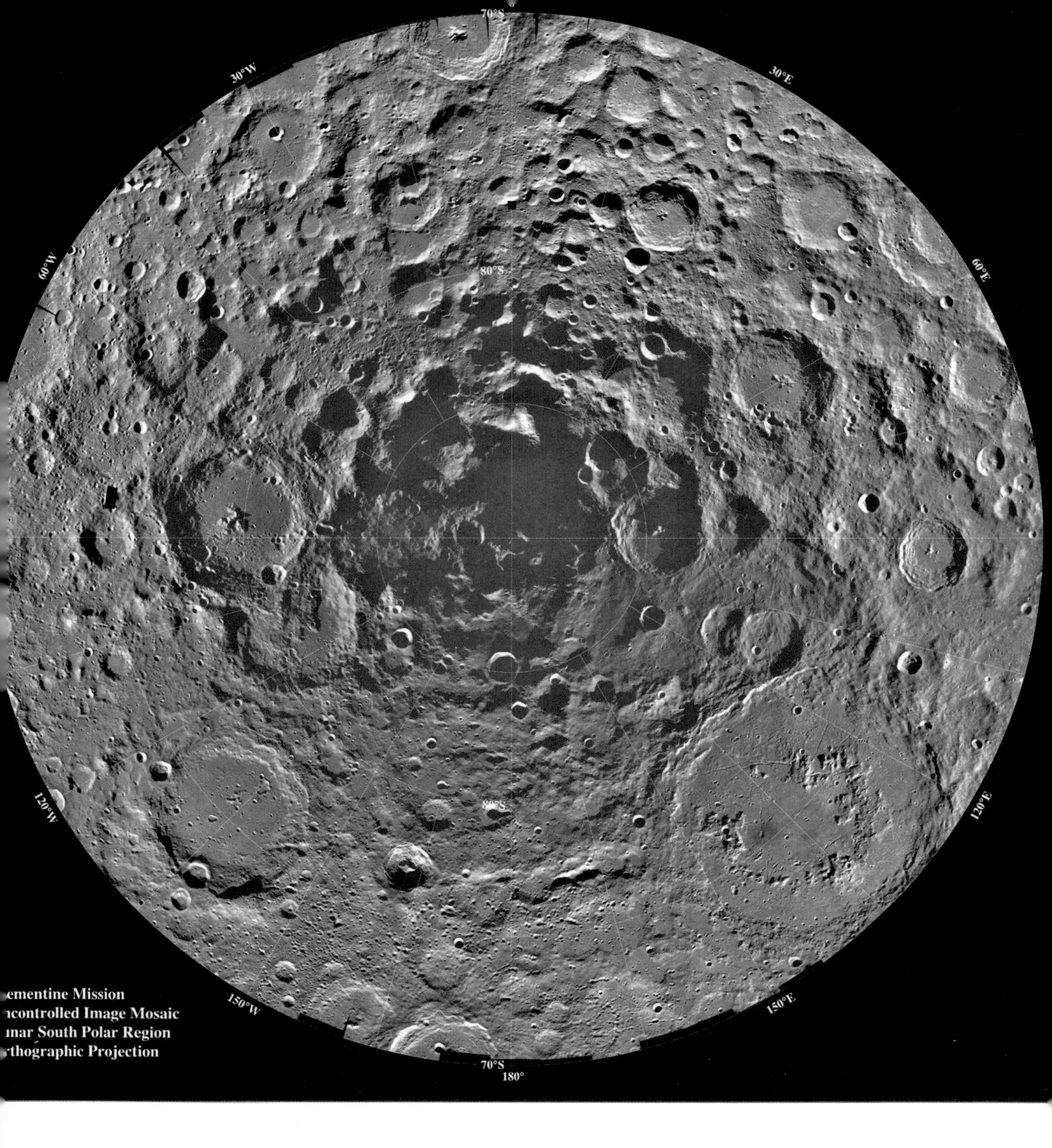

the surface. The north pole *(facing page)* is at a latitude 90 degrees north. This is the same as the north pole designation on Earth.

The lunar south pole *(above)* is of special interest for exploration because more of its area remains out of sunlight, especially in some of the craters. This increases the likelihood of frozen water.

Image credit: NASA

Lunar Seas

Lunar maria (singular mare) are the large basaltic plains that were formed with lava flow by volcanic activity long ago. Viewed from Earth, early astronomers thought they looked like seas.

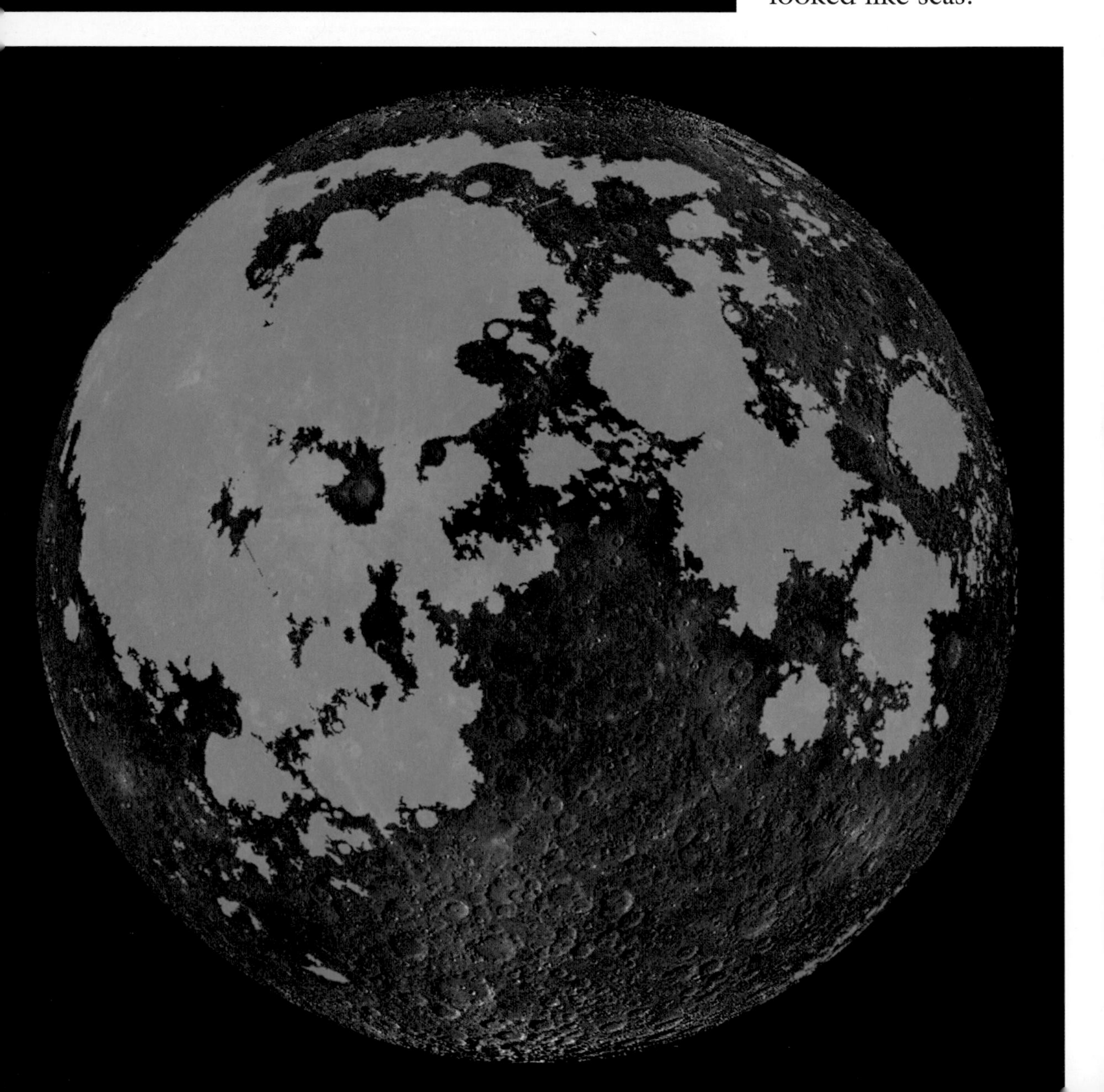

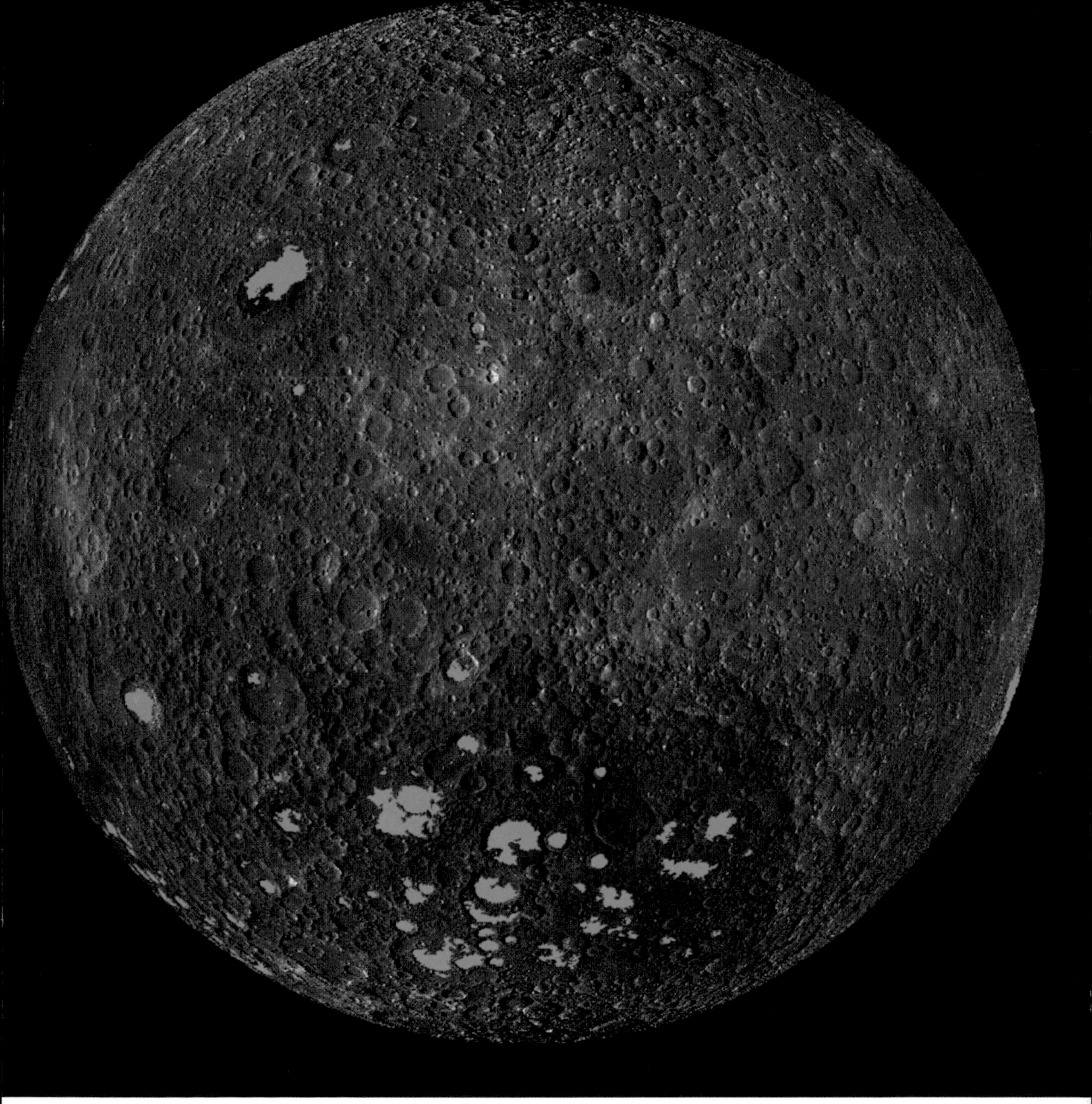

Image set credit: NASA

Maria

This illustration *(facing page top)* indicates the maria on the near side of the Moon. Over 30 percent of the near side lunar surface *(facing page top and bottom)* is covered with maria. On the far side of the Moon *(above)*, there is much less at about 2 percent coverage.

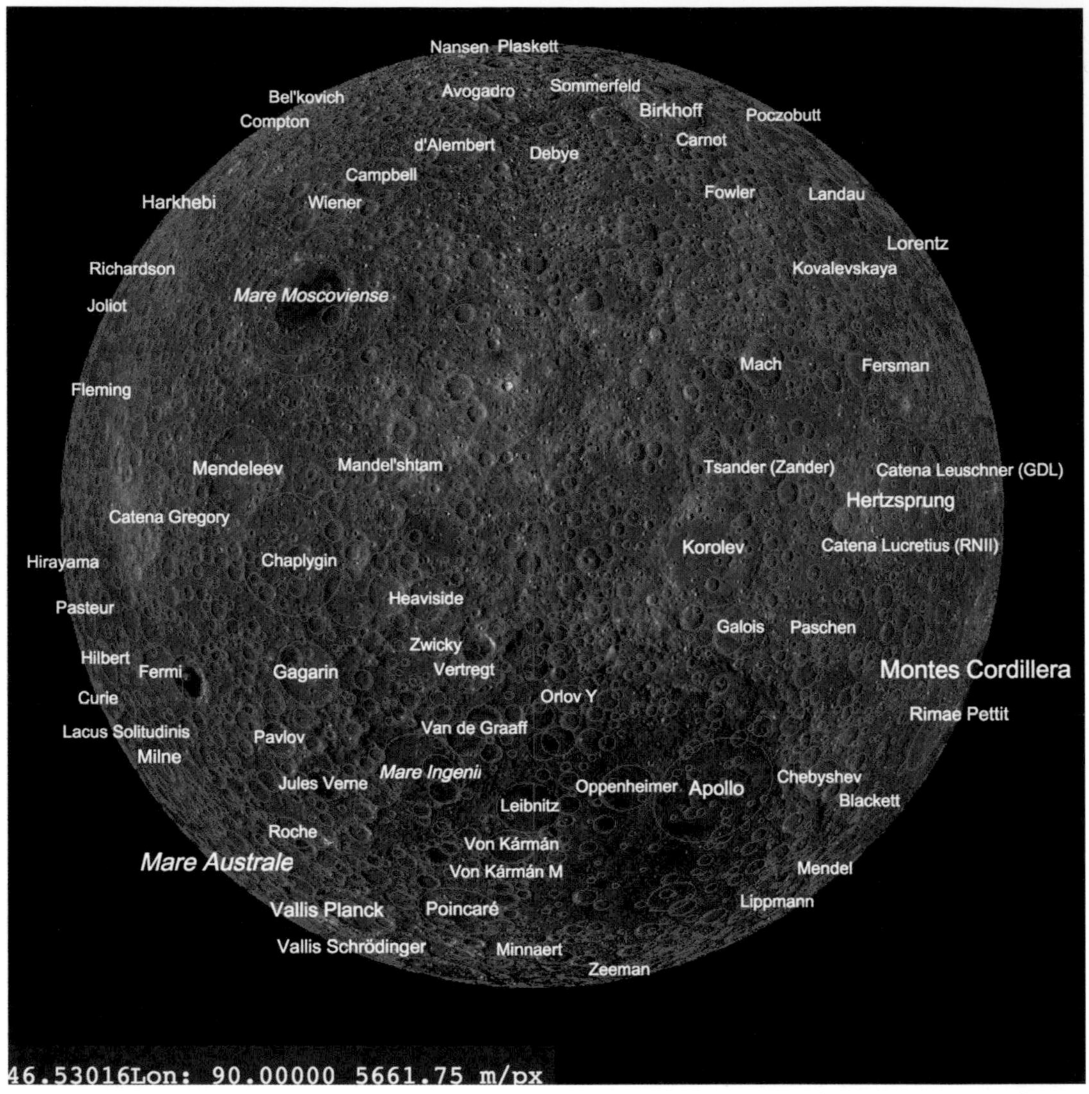

Image credit *(above and facing page)*: NASA

Craters

Most of the Moon craters were formed at a time when there were many impacts in the Earth-Moon system compared to present day. Specimens brought back by the Apollo missions suggest craters are older than 2 billion years. Scientists suggest this was an era of larger projectiles in contrast to smaller, more recent ones.

Craters are formed by mete-oric impact, while larger craters

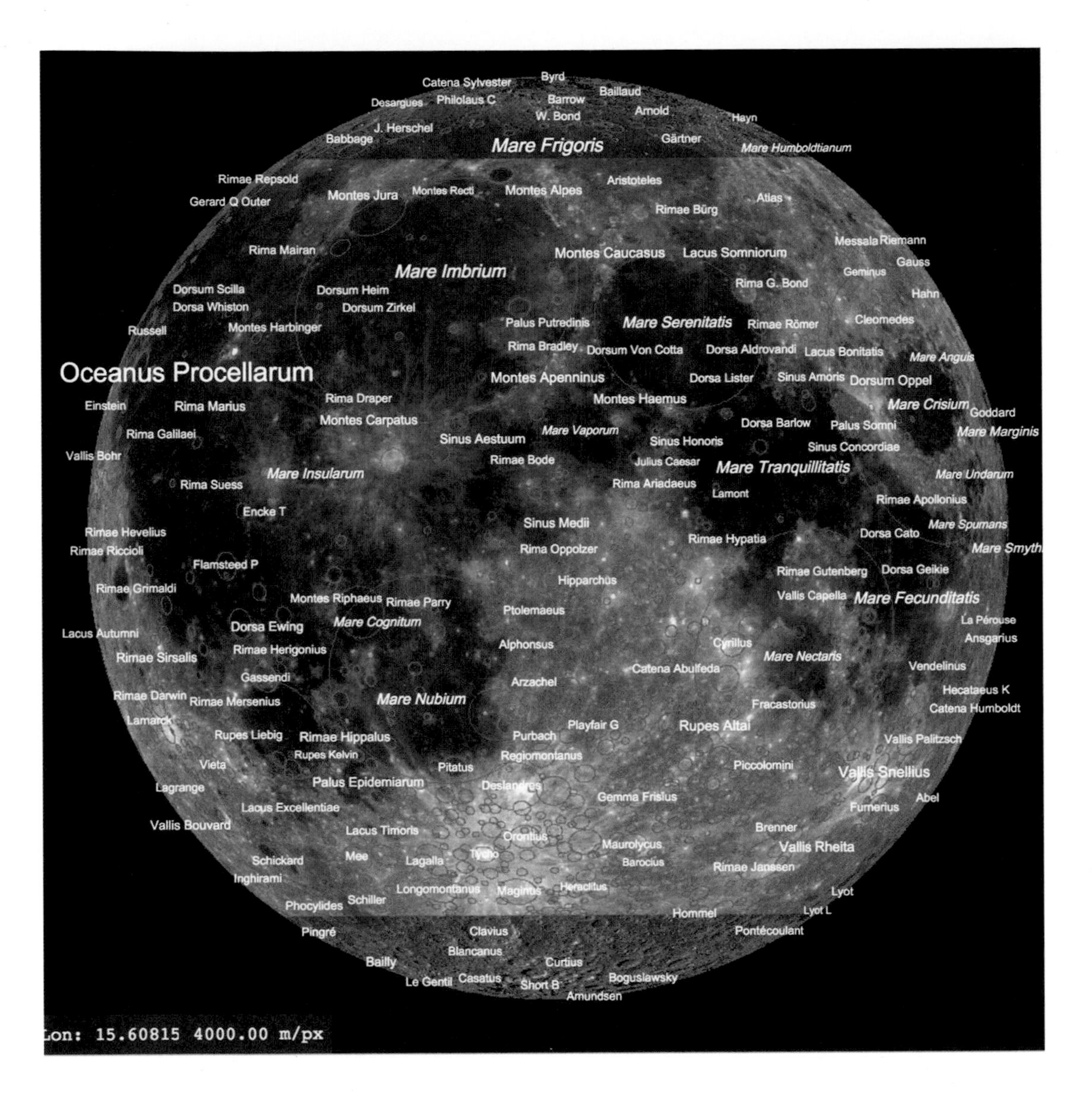

are made by asteroid impact. This gives us insight into a time period when the Earth was likewise bombarded. Today the Earth is protected by its atmosphere, where most small meteors disintegrate or burn up on entry.

Note the differences in the number of craters between the near side and the far side of the Moon. Probably both sides received the same amount of impacts; however, the near side has been resurfaced by the lava flows that created the Maria. This close-up image *(below)* is of the interior wall of the Clerke crater.

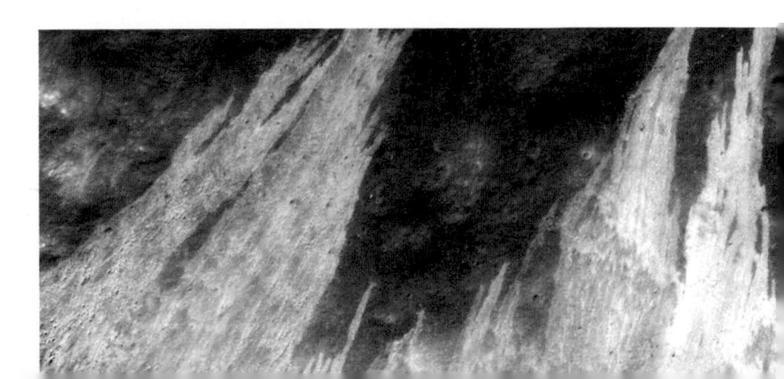

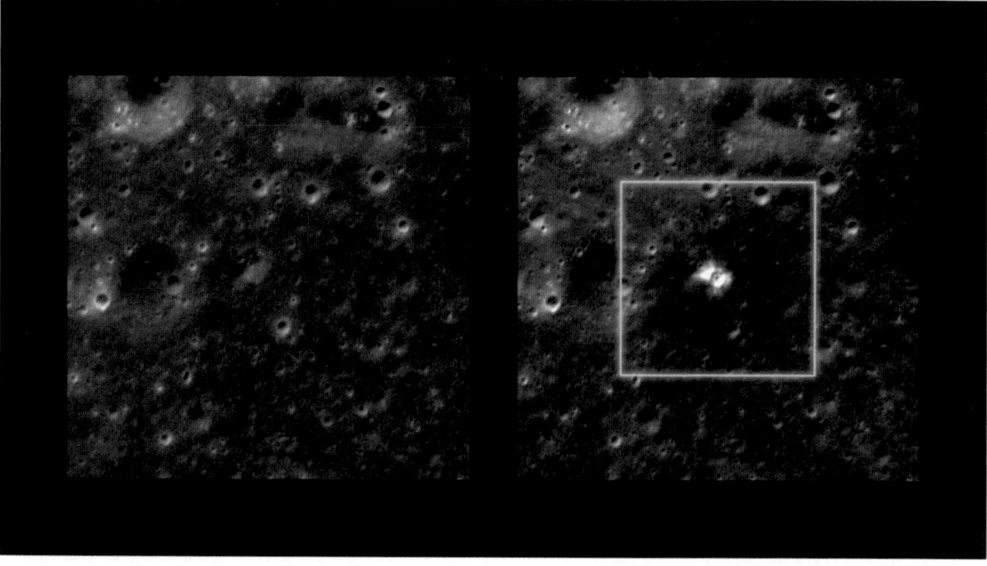

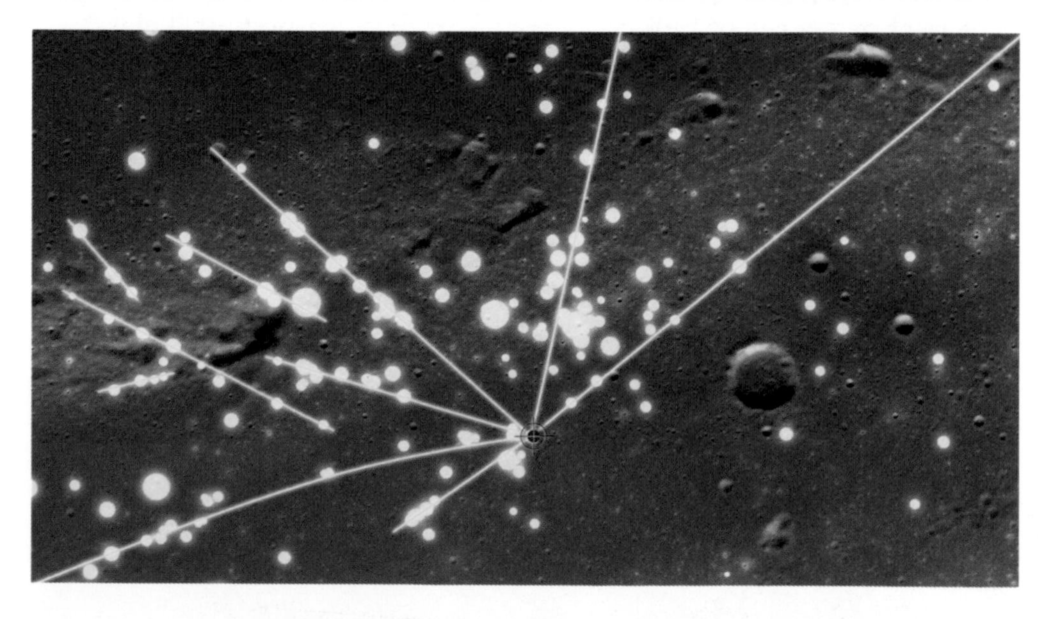

Image set credit: NASA

New Impacts Make New Craters

Beginning in 2009, NASA's Lunar Reconnaissance Orbiter (LRO) started collecting high-resolution images. New images were compared with old images for differences. Over two dozen new craters were found *(facing page top)*. The center image shows a set of before-and-after images. The new crater, formed March 17, 2013 *(facing page center right)* by meteor and is 18 meters wide. The LRO images show more than just the crater. They show the ejected material as well. The illustration *(facing page bottom)* shows the ejected material as connected dots.

The study of craters and their ejected material is important for understanding the Moon's surface environment and for managing the risks for human presence on Earth's natural satellite. Craters are identified, named, and classified *(top)*. This close-up image *(bottom)* is of the eastern rim of Dionysius crater.

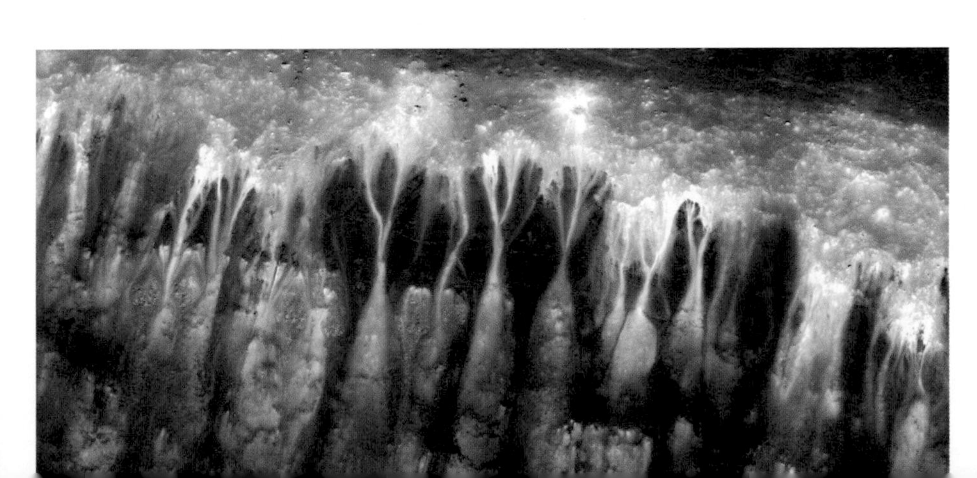

Image credit: NASA

Earth's Craters

Craters on Earth are subject to extreme weathering. Very few impact craters are as well formed as the Pingualuit Crater pictured here. With time, craters on Earth disappear. By studying the Moon's craters, which are subject to less weathering, we can get a better understanding of the Earth's past. There was a time that the Earth experienced the same increase in frequency of asteroid impacts that

Image credit: NASA

the Moon underwent at around 290 million years ago. With the knowledge we gain about the Moon's crater history, the more we will understand the geologic history of the planet we live on.

Key to the Beginning

The Moon could be the key to understanding the beginning of the Solar System. Studying the Moon is like having a fossil record with which to study the Solar System's formation. Some sections of craters have remained out of the reach of solar weathering. These areas will be of particular interest to understanding the early solar system.

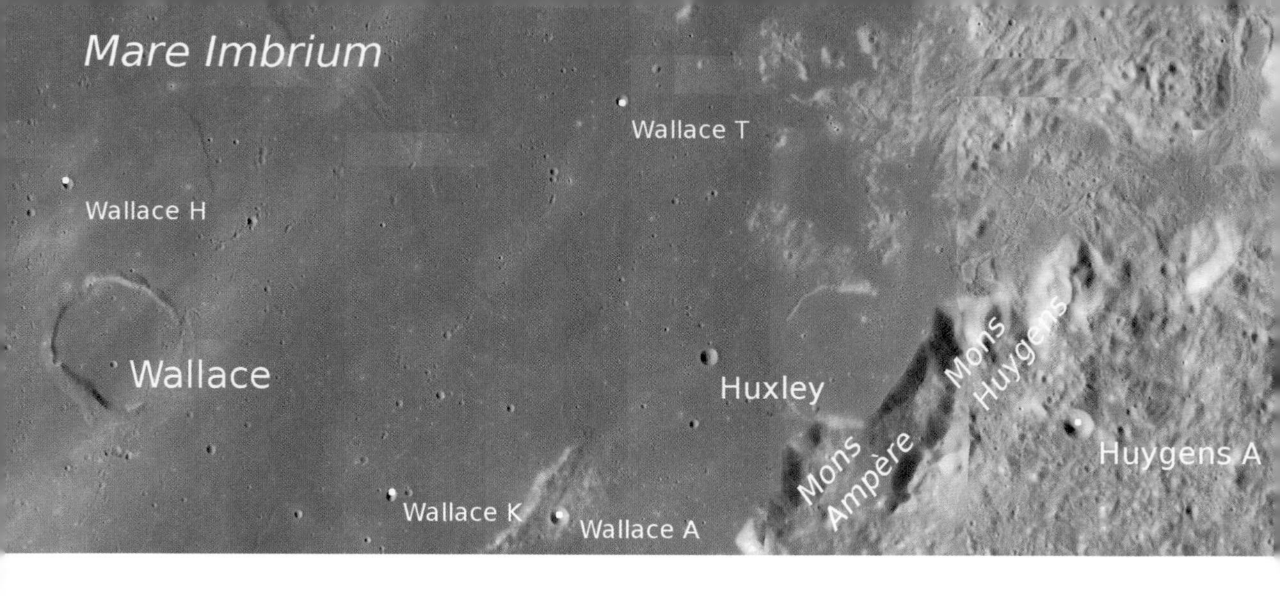

Planetary Nomenclature

Planetary nomenclature is a way to identify planetary and natural satellite features by giving names to the feature and a specific instance of the feature that makes it clear and distinct. It allows the features to be located, described, studied, discussed, and written about.

The task of the International Astronomical Union (IAU) founded in 1919, is to select official names for features of Solar System bodies. The IAU has a group called the Working Group for Planetary System Nomenclature (WGPSN), who meet every three years and undertake this task.

Names should be based on earlier nomenclature, equitable name choices from international ethnic groups, countries, and gender on each map, with some exceptions. Names with political, military, or modern religious significance should not be used. The exception

is political figures from before the 19th century. Official names are not given to newly discovered satellites until their orbital elements or features are known.

The word "lunar" is used as an adjective to describe anything associated with Earth's natural satellite—the Moon. Per NASA, when referring to the Earth's natural satellite, the word "Moon" is capitalized because it is a proper place name. Natural satellites of other planets are referred to as "moon" without capitalization. Lunar feature descriptors that are often included as part of, preceding, or following a specific place name are: crater (circular depression—impact or volcanic), lacus (small plain—lake-like appearance), mare or maria (smooth plain), paludes (low plain), sinus (small plain), mons (mountain), montes (mountain range), rupes (an escarpment), and valles (valley).

Lunar Mountains

The Moon has 48 mountains with official names. The term mons is used to describe mountains of any geological structure and origin on a celestial body. They usually result from tectonic, impact, or volcanic processes.

Mons on the Moon are often named after a nearby crater, such as Hadley C crater and Mons Hadley *(facing page bottom)*. Its summit is estimated to be about 13,000 feet (9407 meters) high.

Montes Caucasus *(below)* is a mountain range. The range was named after the Caucasus Mountain on Earth. These images of crater, mountain, and mountain range were captured during the Apollo 15 mission, in July of 1971.

Image set credit: NASA

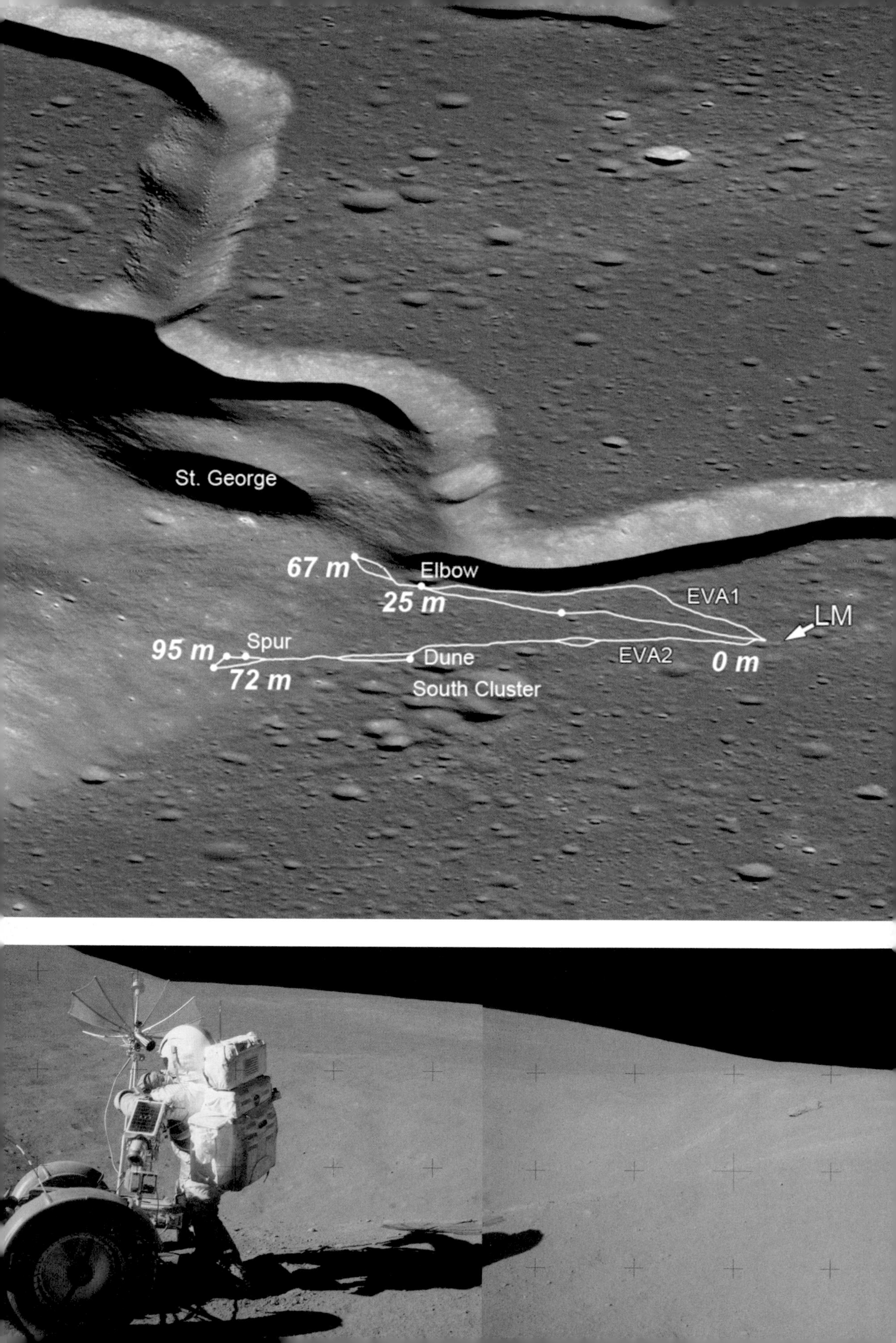
St. George
67 m
Elbow
25 m
EVA1
LM
95 m
Spur
Dune
EVA2
0 m
72 m
South Cluster

Tracing Human Exploration

Humans have still not quite reached every corner of the Earth. Some places are difficult to explore, like the bottom of the oceans and the tops of mountains. Other places, like rivers, have made exploration and eventual settlement easier. There are many reasons why some locations are chosen and not others: access, safety, communications, water, health, and building resources.

Moon exploration will be no different. Safety, survival, travel, and local resources will be necessary components. The only crewed landings, the Apollo missions, needed to be on the near side of the Moon and on a relatively flat area where the rover could drive unimpeded by boulder fields. Geography determines how exploration is carried out. Apollo 15 went out on two EVAs (extra-vehicular activities), seen traced on the image of the Moon *(facing page)*. The first was just under three miles. On the second, they traveled slightly farther.

Image set credit: NASA

Lunar Geology

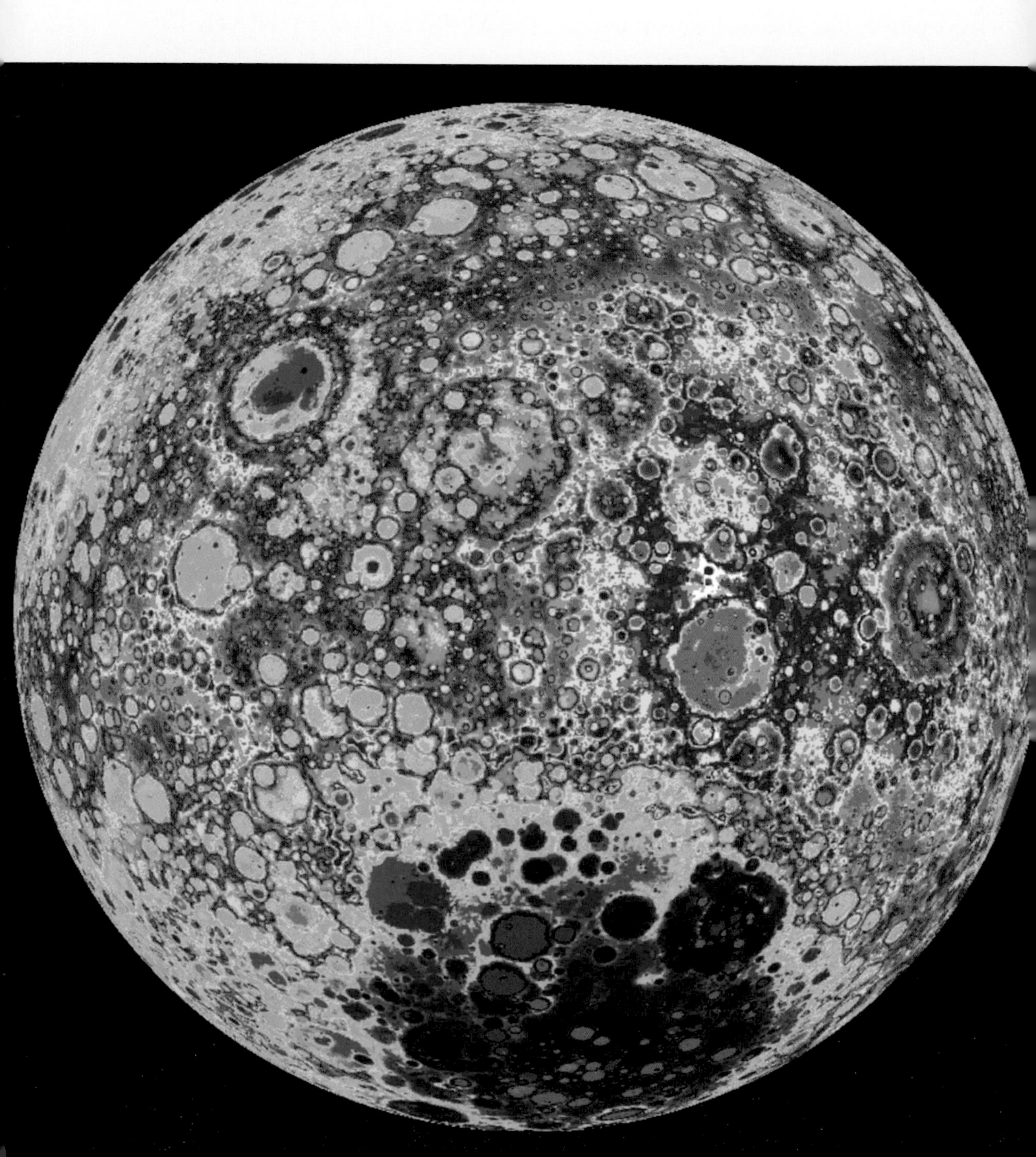

Geology of the Moon

Geography is the study of physical features of a planet or natural satellite, including land use—the arrangement of places and physical features. It includes human activity, such as settlements, population distribution, naming, location of valued resources, exploration, and more. Many geographical studies focus on the physical features and their value to, or history with, humans. The geography of the Moon will be of even greater importance as it becomes even more important to humanity's upcoming explorations in space.

On the other hand, geology is a science that is concerned with the physical structures, substances, and processes that operate on them. The historical context of geological study more often than not predates humanity, although sometimes human factors are very relevant. An astrobiologist searching for life in the form of fossil evidence needs a practical geology background.

Related Fields

Geophysics is concerned with physical properties and processes of the Earth. The same scientific methods and knowledge used in geophysics are also used to study the Moon, planets, and other natural satellites. Other fields are: physics, geodynamics, planetary geology, geochemistry, cosmochemistry, space physics, geomorphology, mineral physics, tectonophysics, geophysical surveying, geodynamics, and of historical importance, selenography. Geology is the general term used here, even though many of the scientific discoveries came from related and more focused fields of study. For example, micromorphology is used to study regolith—the loose rock and dust that is uppermost on the surface of satellites and planets. Through these studies, future visits to the Moon plan to use regolith as a building resource for structures.

Image set credit: NASA

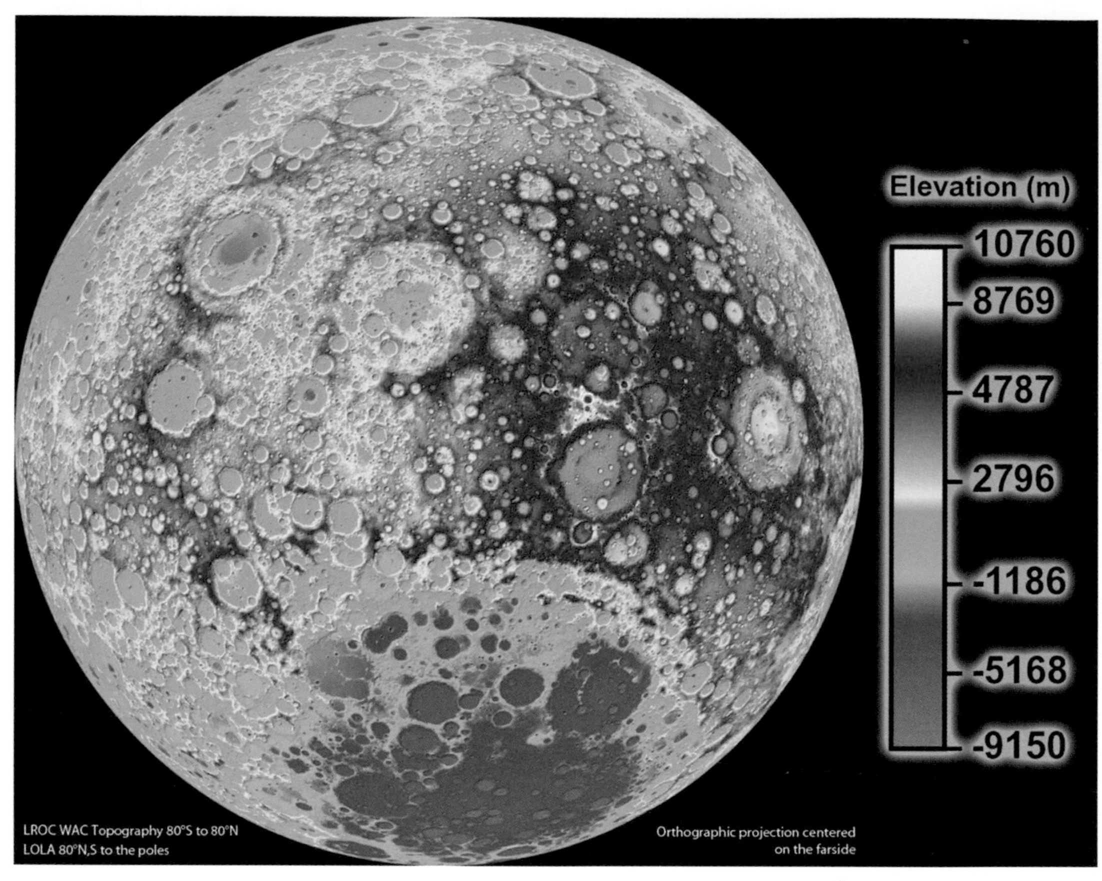

Image Credits: NASA/Goddard Space Flight Center

A Global Map of the Moon's Topography

This image made in 2016 was one of the first high-resolution topography images of the Moon. It was recorded with NASA's Lunar Reconnaissance Orbiter (LRO), using its wide-angle camera and the Lunar Orbiter Laser Altimeter instrument. This is the first accurate high-resolution portrayal of the shape of the entire moon.

Little is known about the morphology of the Moon—the relationships between features, forms, and their structures. Information provided by these instruments is invaluable to geological explorations of the Moon.

Lunar Gravity Map

On the Moon things weigh 16.6% or 1/6 less than they would weigh on Earth. For example, a person weighing a hundred pounds on Earth would weigh and feel like just under 17 pounds on the Moon. This is the reason the Apollo astronauts who walked decades ago on the Moon had that distinctly springy gait.

The weight of objects is caused by the force of gravitation. The greater the mass, the greater the gravitational force. Local variations can be measured using sensitive equipment. In this image, the Moon's mass was mapped using NASA's Gravity Recovery and Interior Laboratory (GRAIL) in 2012. Red illustrates areas of higher mass and blue illustrates areas of lower mass.

Porosity of the Surface

This image shows the porosity of an area of the lunar crust. Data was used from NASA's GRAIL mission, orbital remote-sensing data, and samples that had been collected from NASA's Apollo missions.

Red shows higher than average porosities. Blue shows lower than average porosities. The other colored areas fall somewhere between high and low porosities. White shows areas that were not analyzed.

Scientists say the porosity of this area of crust is a consequence of fractures generated by impact craters over billions of years. The inside of many impact craters is less porous than outside of the craters. This is because temperatures were high enough to melt the materials on impact. In contrast, just outside of the crater, the porosity is higher due to fracturing from the shock waves from impact.

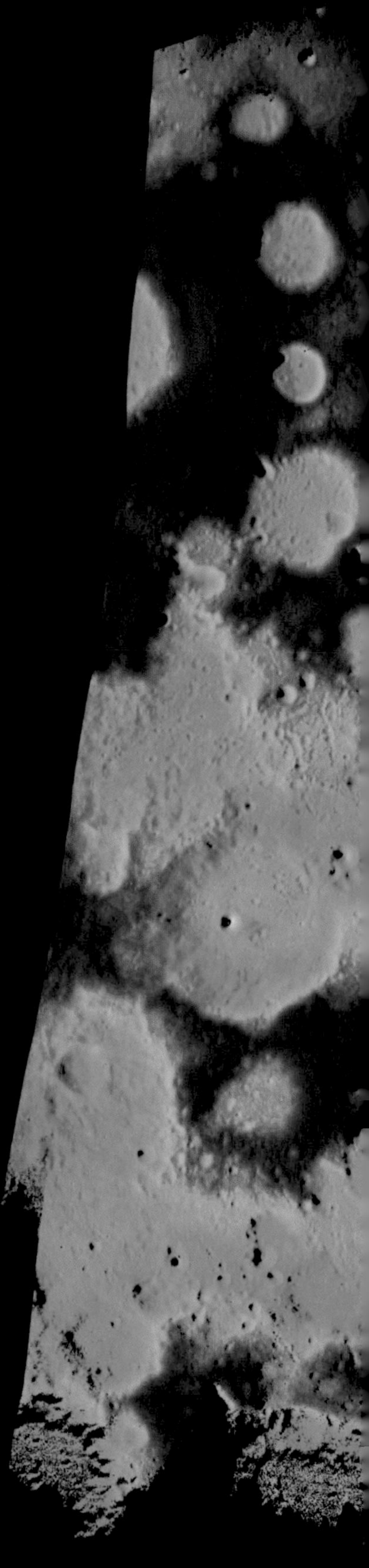

Image credit: NASA/JPL-Caltech/ IPGP

Colorbar for hgt
-4000 -2000 0 2000 4000 6000

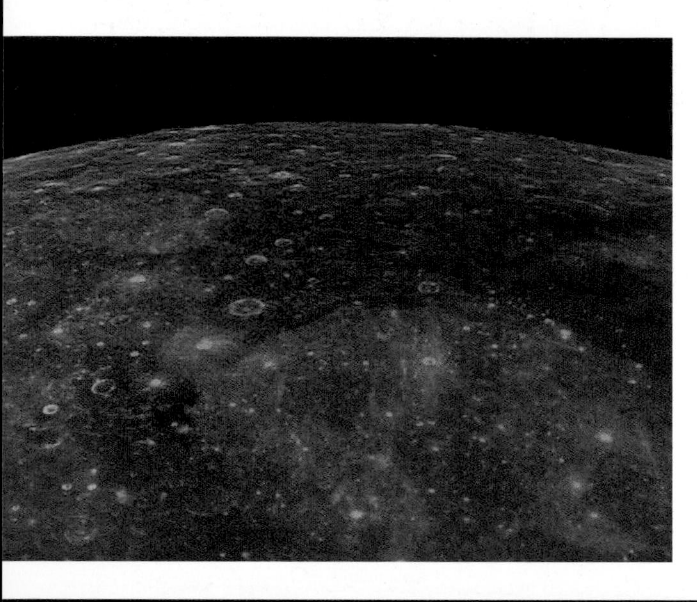

Image credits *(top, left):* NASA/JPL
Image credit *(top, right):* NASA
Image credits *(bottom):* NASA/JPL

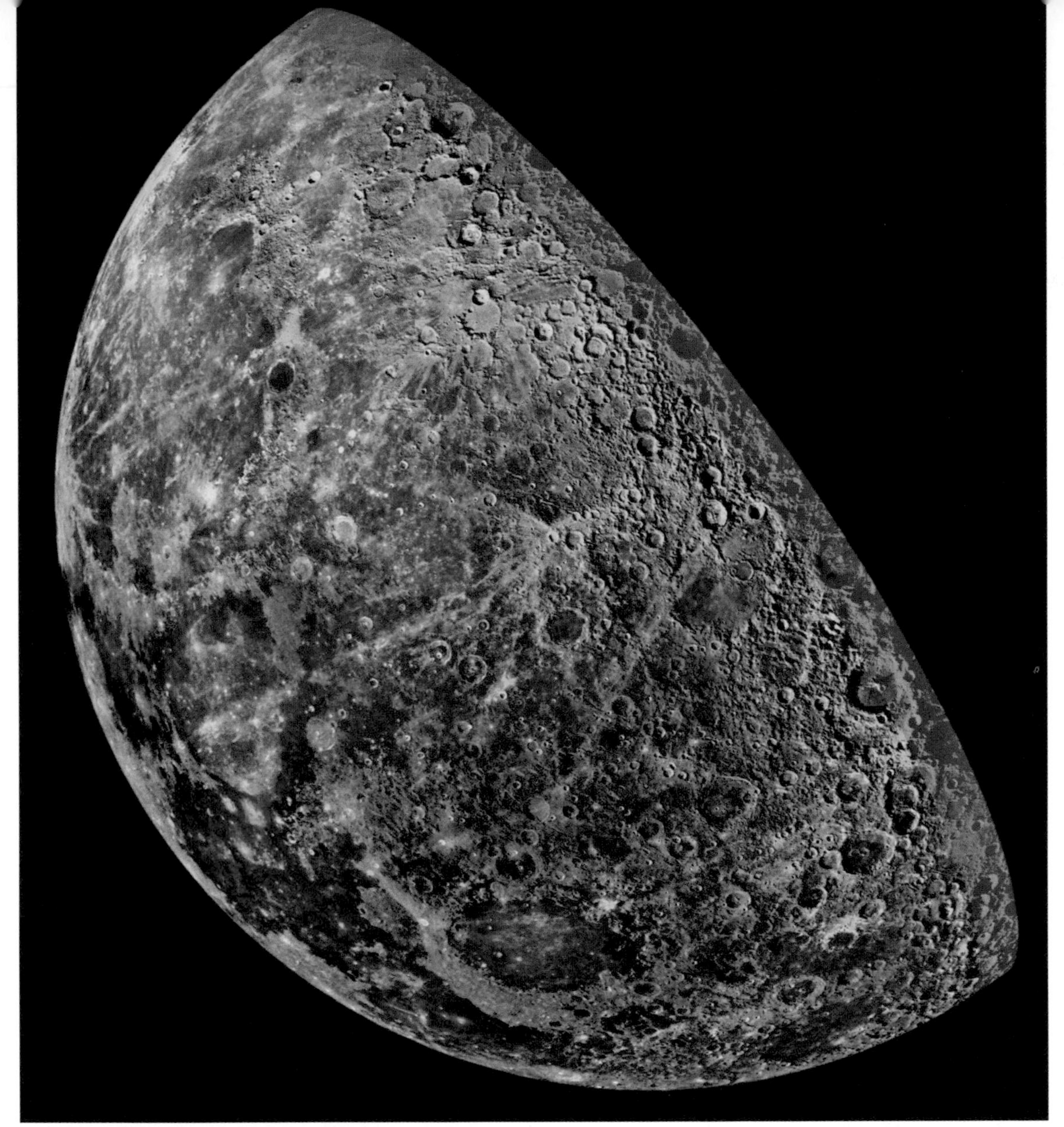

Image credit: NASA

Galileo Spacecraft Images

NASA's Galileo was an uncrewed spacecraft headed to study Jupiter's planet-moon system. It was launched in 1989 by Space Shuttle Atlantis and contained a variety of instrumentation aboard. These four images of the Moon are made from data acquired by Galileo as it left the Earth-Moon system. Galileo's Near-Infrared Mapping Spectrometer was used in a December 1992, Earth/Moon flyby.

Image credits *(top):* NASA/Gateway to Astronaut Photography of Earth
Image credit*(bottom):* NASA

Atmosphere

Atmosphere is the layers of gases that surround a planet or other body such as a natural satellite like a moon. A body will retain an atmosphere if it has a higher amount of gravity and the atmosphere's

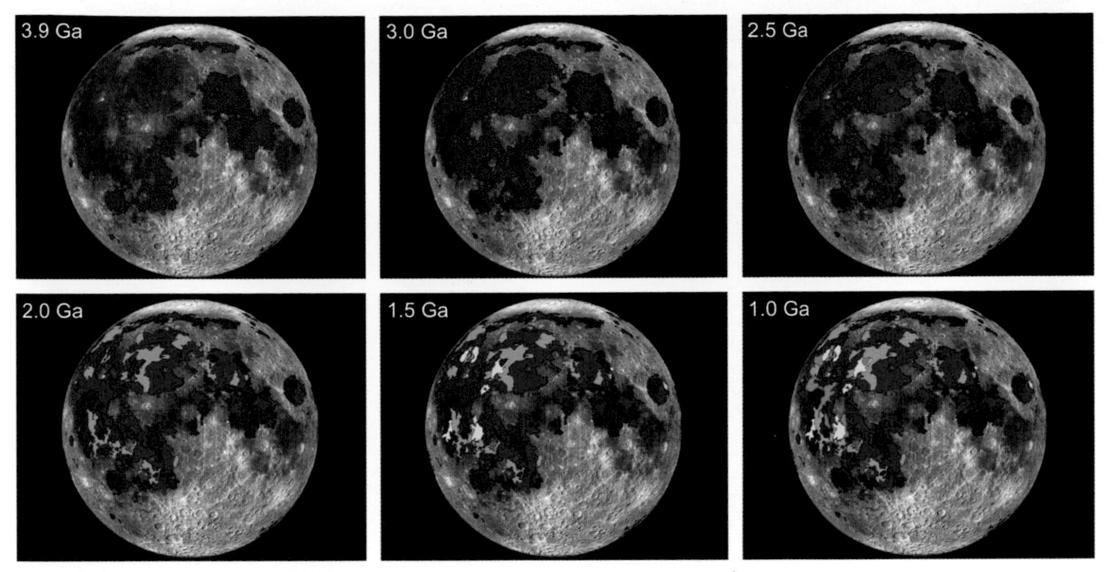

Image credits: NASA/MSFC/Debra Needham; Lunar and Planetary Science Institute/David Kring

temperature is low. A magnetosphere, like the Earth has *(facing page top)*, will help retain an atmosphere by deflecting the destructive solar wind and cosmic rays. Mars *(facing page bottom)* at one time had a more significant magnetosphere that protected a more impressive atmosphere than it has today. It is theorized that the Moon had a magnetic field that produced a magnetosphere that allowed an atmosphere to exist for a period of time. A planet's rotation and its orbit contributes to kinetic energy that could create a liquid state in the Moon's core. This would help to make a magnetosphere that could deflect solar wind and cosmic rays and produce an atmosphere.

Lava Indicates Former Lunar Atmosphere

The Moon had an active volcanic period in its history *(top)*, 3.9 to 1 billion years ago. Most of the mares face Earth. They appear as darker areas and are lava plains. The Apollo crews returned with some of the volcanic rock. Besides producing the iron and magnesium rocks, the volcanic eruptions created a lot of gas which contributed to an atmosphere that lasted about 70 million years. The atmosphere was eventually lost to space or incorporated into the Moon's surface.

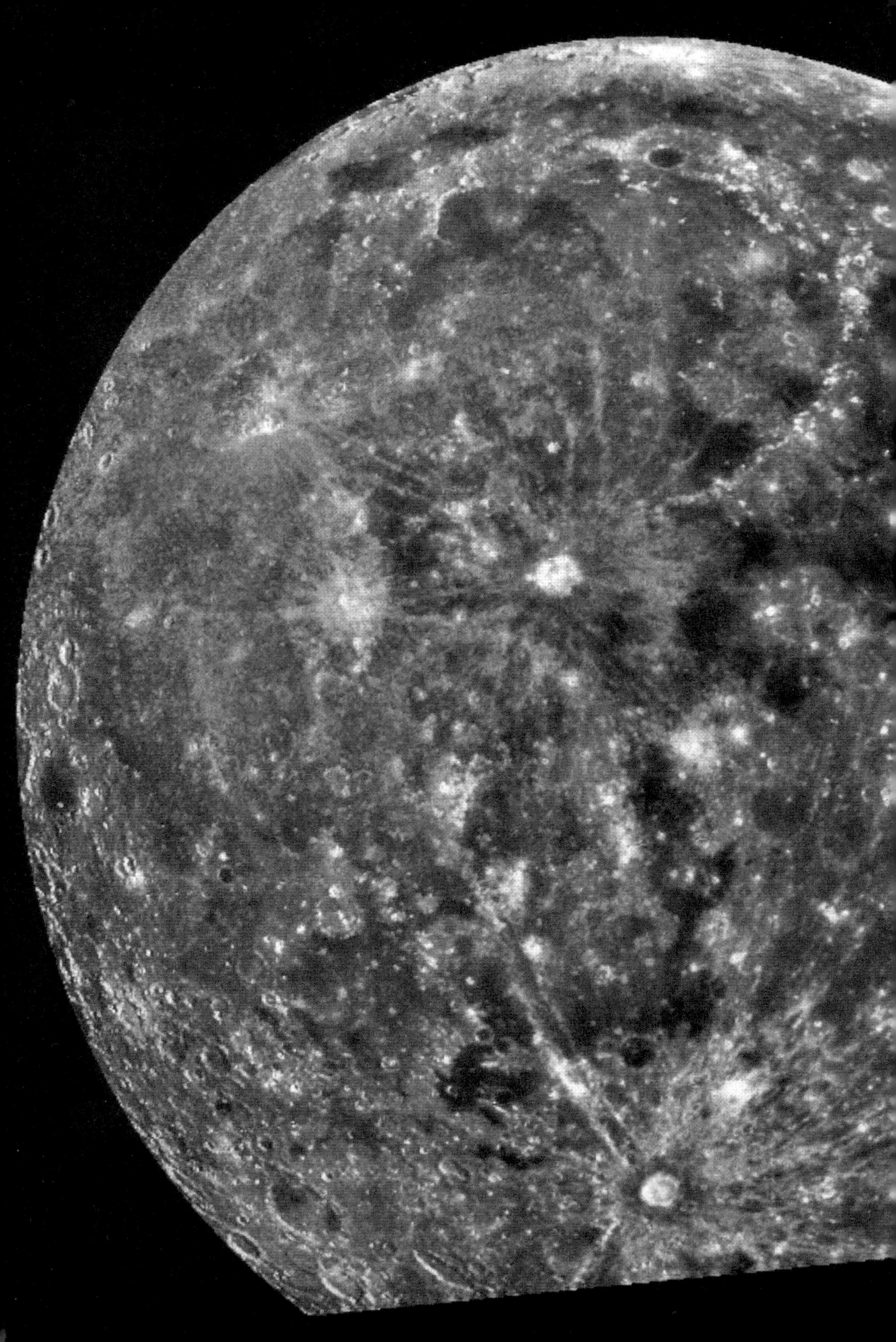

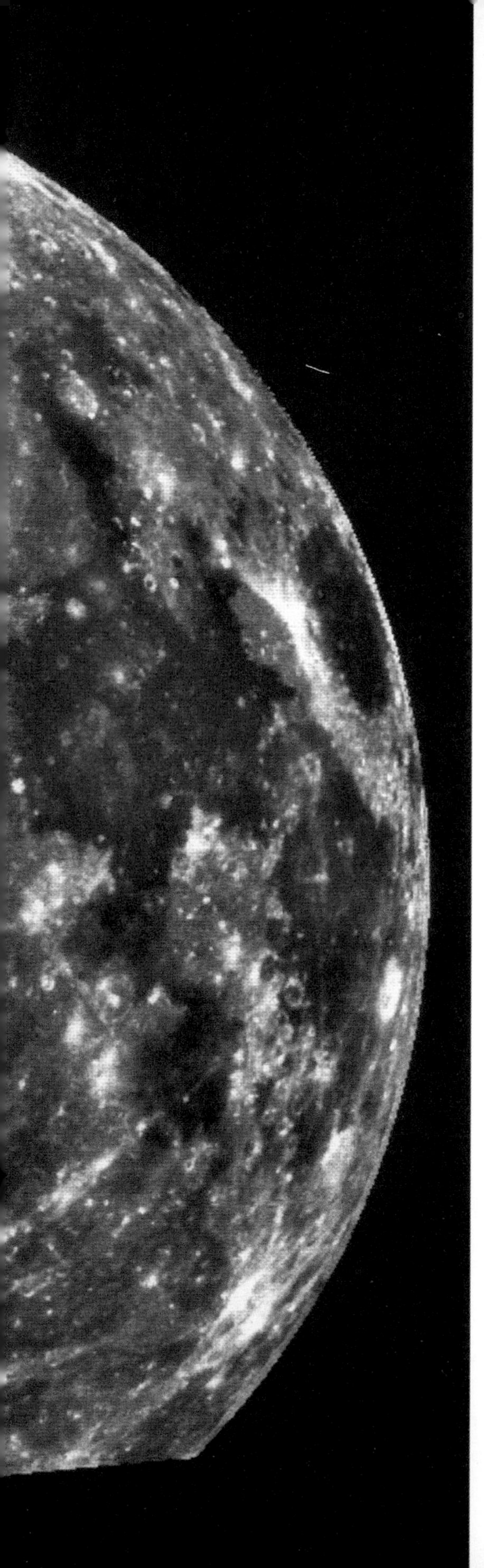

Ancient Volcanic Lava Flow

Today, the near side of the Moon that faces Earth holds the majority of the mares containing the lava plains.

This false-color processed image helps to understand the surface soil composition. According to NASA scientists, the red/pink areas are the highlands. The blue and orange are the ancient volcanic flows. The blue areas have more titanium than the orange areas. The titanium is in the form of ilmenite, which could one day be separated out as a source of oxygen, iron, and titanium.

The majority of the volcanic eruptions occurred between 3 and 3.5 billion years ago. Although, some samples show some eruptions were as old as 4.2 billion years ago and some as recent as 1 billion years ago.

Image credit: NASA

Today's Atmosphere on the Moon

In practical terms, the Moon is in a vacuum and does not have an atmosphere. The atmosphere on the Moon would be comparable to the density of the atmosphere that exists where the International Space Station orbits the Earth. Drawings *(facing page)* from the Apollo missions illustrate the visual appearance of the lunar atmosphere. The thin lunar exosphere at times has this appearance because of lunar crustal magnetic fields reflecting the solar wind protons in a backstream. The Moon does not have a magnetosphere like the Earth that can protect its atmosphere, but it does have magnetic anomalies that affect how the solar winds react around it.

Image set credit: NASA

Sensitive instruments such as the Lunar Atmospheric Composition Experiment (LACE) *(facing page)* have detected small amounts of helium, argon, and possibly neon, ammonia, methane, and carbon dioxide. This is not the right atmosphere to support humans because we would need pressurized oxygen to survive. Living and inside work accommodations would need to be pressurized with oxygen mixtures. Spacesuits would need to be worn outside for work and travel, and oxygen supplies carried. Living accommodations and suits would need to protect us from space radiation, temperature extremes, and pressure.

Some gases that make up the Moon's atmosphere are released as a result of radioactive decay inside the Moon. Radon and helium are released. Another contribution to the lunar atmosphere is through sputtering—the bombardment of the surface by solar wind, sunlight, and micrometeorites.

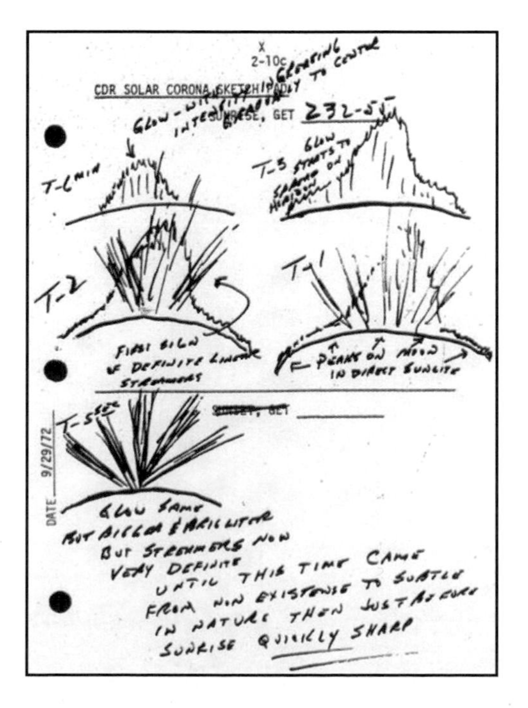

These gases in the lunar atmosphere will do one of several things. The Moon's gravity can incorporate the gas into the regolith—the loose rock and dust on the Moon's surface. Or, the gas can escape the Moon entirely, ionized and carried away in the solar wind magnetic field, or lost into space.

Unlike the Earth's atmosphere, the Moon's minimal atmosphere cannot absorb measurable quantities of radiation, does not appear layered or self-circulating, and requires constant replenishment because a high proportion of its gases are lost to space.

Several methods have been used to discover the composition of the Moon's scant atmosphere. Earth-based spectroscopic methods detected sodium and potassium. The Lunar Prospector Alpha Particle Spectrometer concluded the isotopes radon-222 and polonium-210 were present. Instruments placed by the Apollo astronauts detected argon-40, helium-4, oxygen and/or methane, nitrogen and/or carbon monoxide, and carbon dioxide.

Lunar Reconnaissance Orbiter (LRO)

The Lunar Reconnaissance Orbiter (LRO) *(bottom)* was launched on June 18, 2009. The LRO was to produce a complete detailed high-resolution topographic map of the Moon's surface. Scientists can compare earlier images with more recent ones to record changes.

The instruments on board are as follows:

- Cosmic Ray Telescope for the Effects of Radiation (CRaTER) has a primary goal to characterize the global lunar radiation environment and its biological impacts.

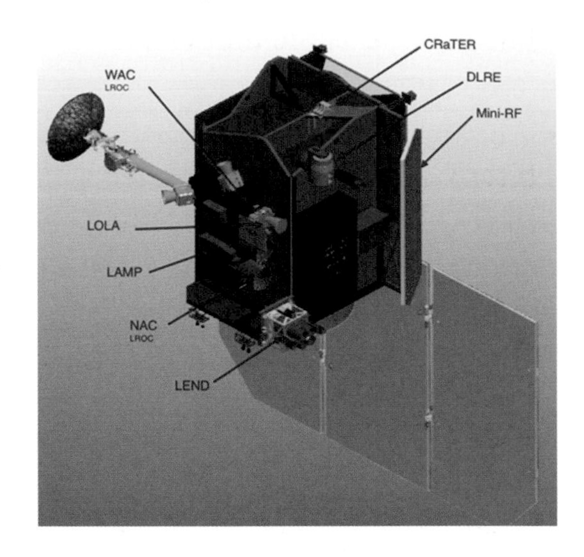

- Lyman-Alpha Mapping Project (LAMP) searches for water ice in craters, using ultraviolet light.
- Lunar Exploration Neutron Detector (LEND) takes measurements, makes maps, and detects possible near-surface water ice.
- Lunar Orbiter Laser Altimeter (LOLA) helps to make a global

lunar topographic model and geodetic grid.

- Lunar Reconnaissance Orbiter Camera (LROC) is actually several high-resolution cameras. One single wide-angle camera (WAC) and two narrow-angle push-broom imaging cameras (NAC) with spectroscopic sensors.
- Miniature Radio Frequency radar has a Synthetic-Aperture Radar (SAR) which uses successive pulses of radio waves and has located potential water-ice.
- The Diviner Lunar Radiometer (DLRE) is an experiment that measures lunar surface thermal emission.

The zigzag appearance of this image *(bottom)* is the result of meteoroid (a small natural object in space) hit on October 13, 2014 that shook the sensitive instruments.

Image set credit: NASA

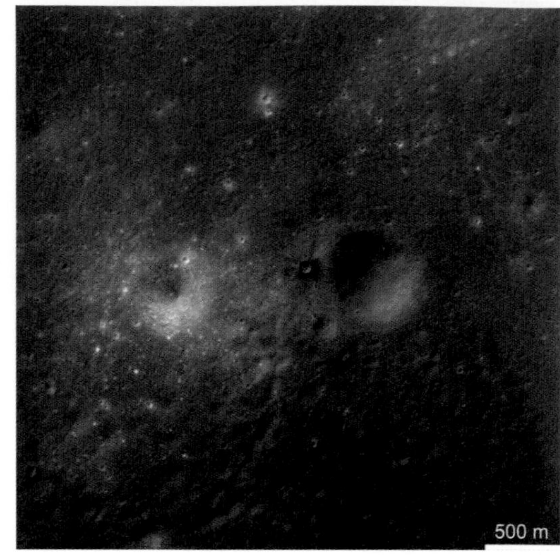

Image set credit: NASA

Dark Ejecta on Crater

Shorty Crater *(top right and closeup, top left)* is just north of the crater Jules Verne. This crater has dark ejecta—material that is forced out of the crater on a meteor's impact. Craters more typically have ejecta that is brighter than the surrounding area because this new material hasn't weathered. A few craters will have dark ejecta because the underlying material has low reflectance material, either rock or regolith. The reason for this crater's dark ring and ejecta is most likely due to pyroclastics—primarily volcanic material—beneath the surface. This could be a lava field such as is found in a mare, or fragments from volcanic activity.

Moon Dike Map

Magmatic dikes *(facing page middle right)* are features that occur when lava forms in fissures of preexisting rock. This one formed on Earth. Many have been mapped on the Moon. The dikes have a higher density than other areas with a resulting increase in gravity. This is important in understanding the geology of the Moon. The increase in gravitation can also affect a spacecraft's orbital performance.

Gravity Recovery and Interior Laboratory (GRAIL) data made the gravity maps *(facing page)* possible. Higher density areas are called mascons which is a combination of the words *mass* and *concentration*.

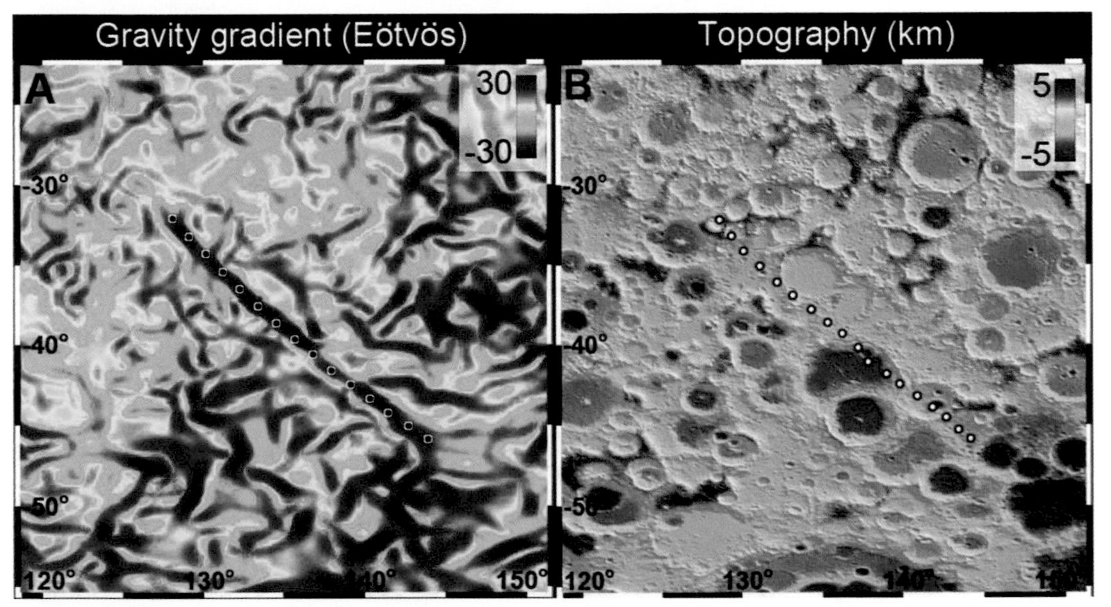

Gravity gradient (Eötvös)
Topography (km)
A
30
-30
B
5
-5
-30°
-40°
-50°
120°
130°
140°
150°

Gravity gradient
(Eotvos)
30
0
-30

Moon Rocks

Moon rocks come from three sources. NASA's crewed Apollo missions to the Moon, three Soviet Luna programme uncrewed probes, and rocks that have fallen to Earth after being ejected from the Moon. These ejected specimens are found on the surface of Earth and are quite rare.

Image set credit: NASA

Apollo Moon Rocks

The Apollo astronauts, over the course of the crewed missions, brought about 840 pounds of rocks and soil back to Earth. The rock *(top left)* is lunar basalt. The rock below brought back by Apollo 17 is described as a vesicular, medium-grained, high-titanium ilmenite basalt, and is 3.7 billion years old. These specimens are still studied today.

Lunar Cross Section

The Moon has a solid iron-rich core that is surrounded by a liquid iron layer. The core is a scant 149 miles (240 kilometers), and the liquid layer shell is 93 miles (150 kilometers) thick. Next, from the core's molten outer layer, the mantle extends outward to the crust. The mantle is made up of minerals that contain magnesium, iron, silicon, and oxygen. The crust is 43 miles (70 kilometers) thick on the near side and twice the thickness on the far side at 93 miles (150 kilometers).

Ancient Volcanos

Most of the lava that flowed on the Moon create the maria. There are some volcanos that created lunar domes *(below)*, which is a type of shield volcano that is sometimes found on Earth, too. A maria is a large, extensive, flat plain. A shield volcano has a low altitude compared to its horizontal span. The flatness of both the maria and the shield dome is due to the highly fluid lava which has a low viscosity. These craters can be less circular than impact craters.

There are no active volcanoes on the Moon. Most of the Moon's volcanism happened over 1 billion year ago. There is evidence that lava did flow as recently as 100 million years ago.

Image credit *(facing page and below)*: NASA

Moon Quakes

The Moon is thought to have no tectonic plates. Moon tectonics relates to the satellite's crust and the processes within it. So the equipment from the Apollo Lunar Surface Experiments Package (ALSEP) pictured here is used to understand moonquakes that are a result of forces other than plate

movement. Most likely these are tidal forces between the Earth and the Moon. The instruments were put in place by the Apollo 12, 14, 15, and 16 missions. They were functioning until 1977.

The experiments followed the propagation of seismic waves through the Moon. These studies gave new details of the Moon's internal structure. Other seismic activity is caused by meteorite impacts, thermal quakes from monthly rotations in and out of sunlight, and lastly, shallow quakes. Shallow quakes are more disruptive and could be an issue for future building and exploration.

Image set credit: NASA

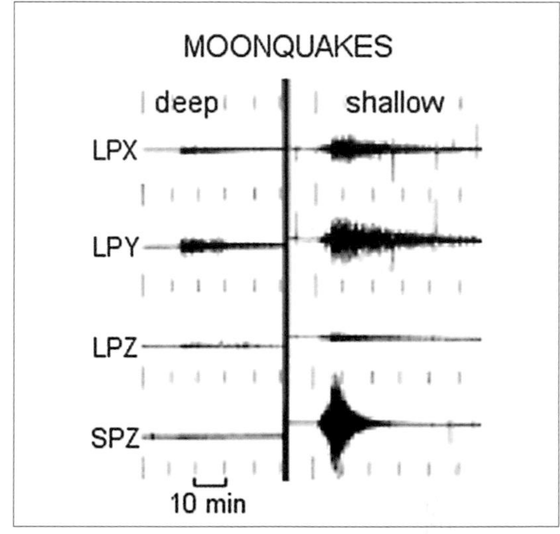

Image set credit: NASA

Regolith

Regolith is dirt, broken rocks, volcanic ash, dust, gravel, pebbles, grit, sand, and mud. It is the layers of loose stuff that accumulates on the rocky foundation of Earth, rocky planets, natural satellites, and some asteroids. The regolith of Mars and the Moon are varied and unique to each.

This footprint *(above)* is from one of the astronauts from the Apollo missions to the Moon. Un-

like the Earth, the regolith on the Moon does not undergo the same weathering due to atmospheric process. On the Moon, mechanical breaking of rocks and minerals *(right)* is called comminution and is carried out by meteorite and micrometeorite impacts. Agglutination is the welding together of particles, such as the glass made by meteorite and micrometeorite impacts. An example is the orangy color of microscopic glass beads *(top left)* made by long ago volcanic processes on regolith, which can be made by solar wind as small particles can be impacted by ions and high energy particles.

Studying regolith is a means to understand the composition and geological processes of the Moon. Regolith may be the main building material for housing and other structures on the Moon and Mars during upcoming explorations.

Moon's Gravity Map

Gravity maps help to understand a planet or its satellite better. They help to tie in what is on the surface with what is below the surface. Below is a gravity map of Mars.

This is a color-coded map *(facing page)* showing the strength of surface gravity around the Orientale basin on the moon. The image was created using data from NASA's GRAIL mission. Its mission was to produce a high-resolution map of the Moon. The image on the facing page is of the Orientale basin. Because the probes were so close to the surface, the measurements were very sensitive.

Image credits: NASA/GSFC/Scientific Visualization Studio

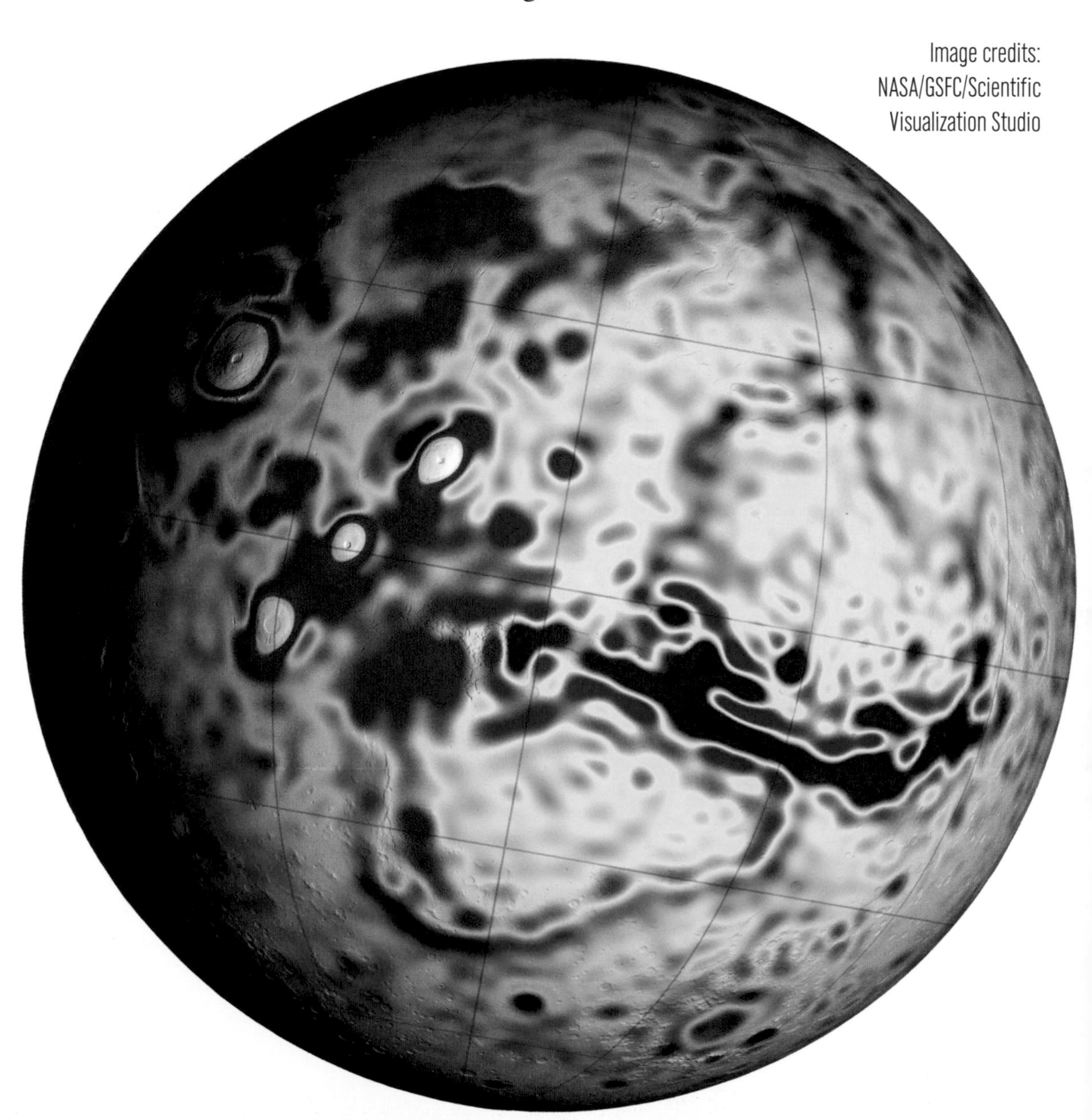

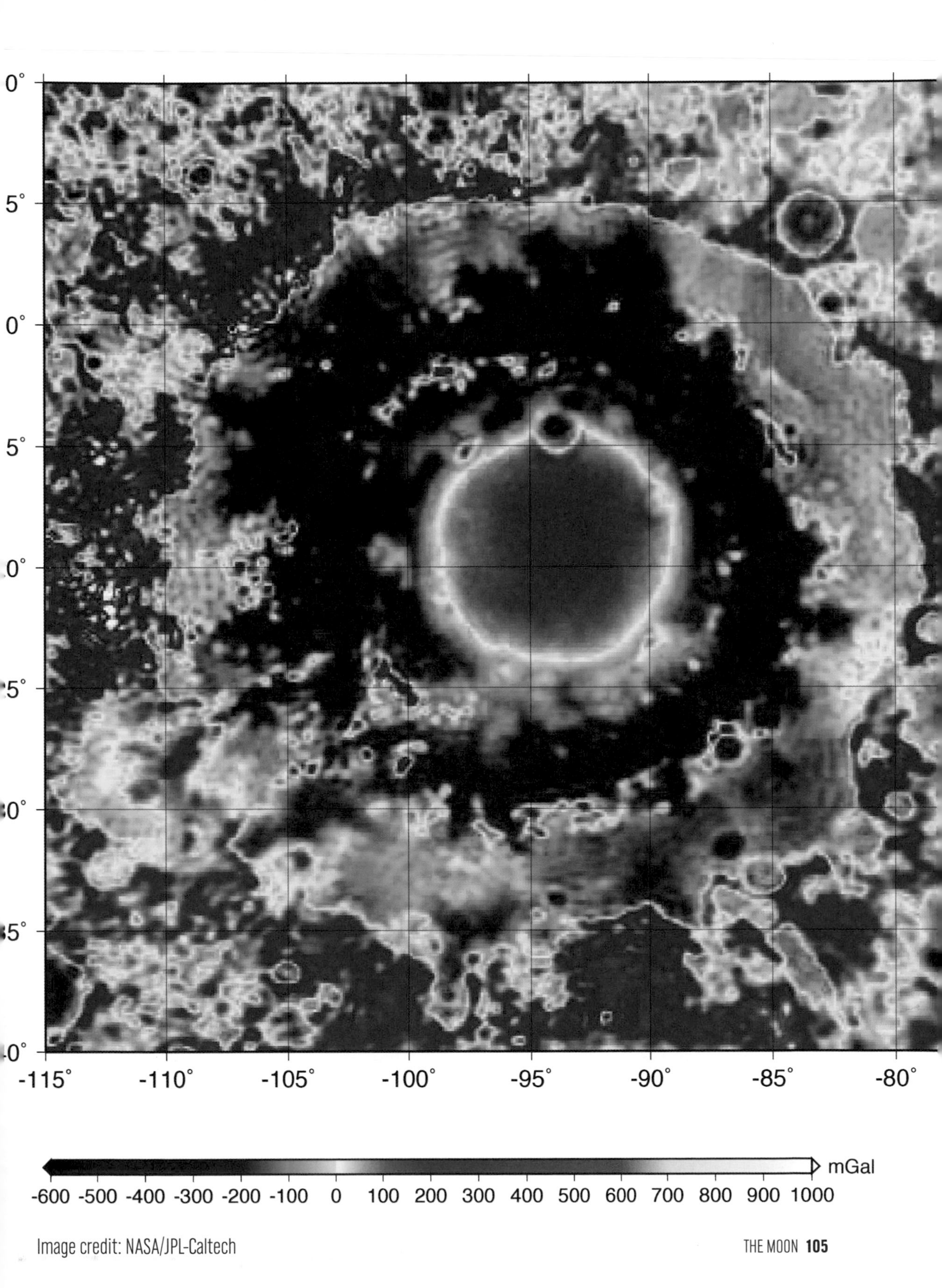

Image credit: NASA/JPL-Caltech

Living on, Working on & Exploring the Moon

Image set credit: NASA

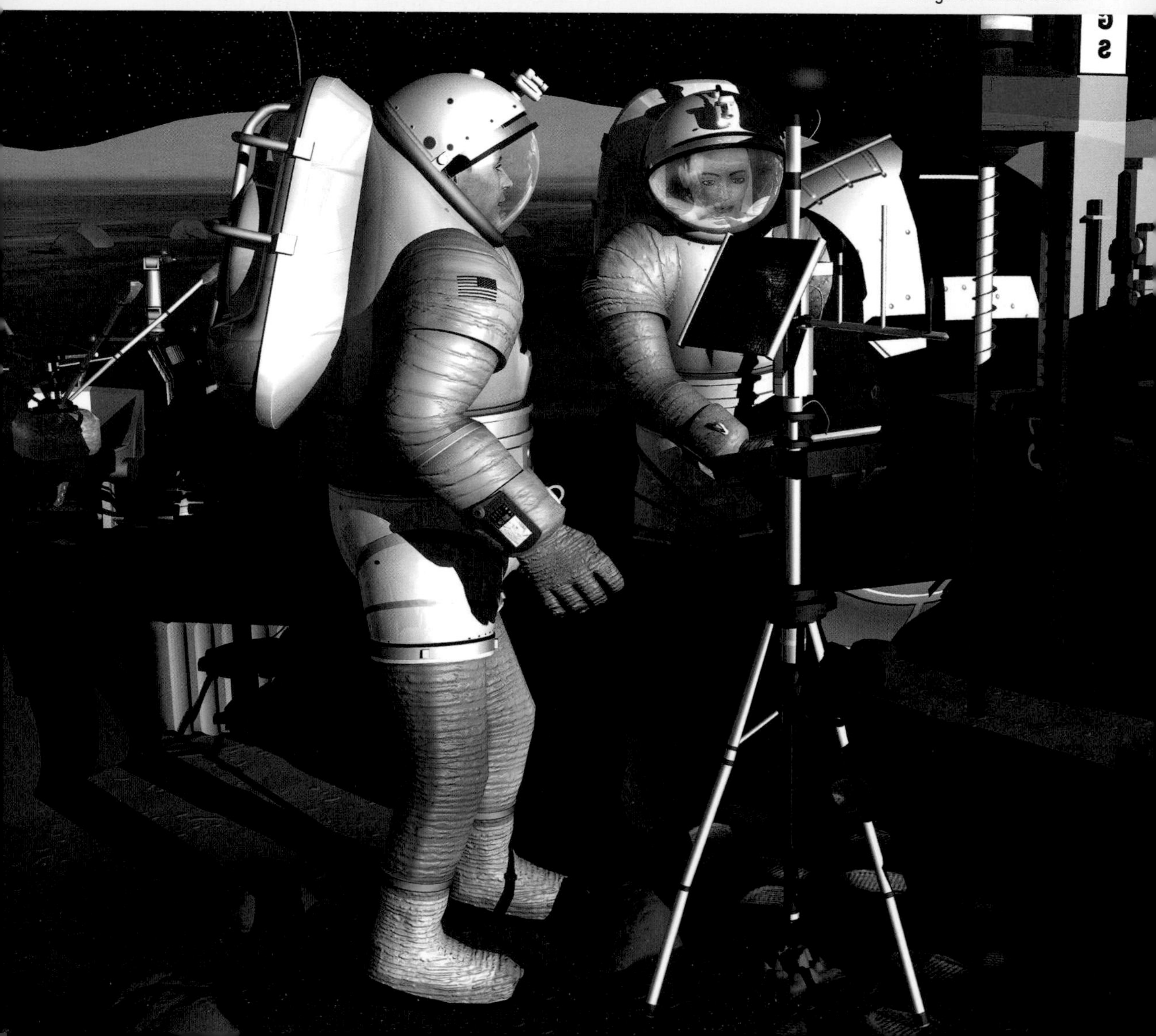

Research for Settlements

Before comprehensive lunar sites can be accomplished, a few hurdles need to be cleared. These obstacles include creating environments that protect from cosmic radiation and other environmental variables, as well as the problem of the high cost of spaceflight. The obstacle of using less from the Earth's surface and using more from the Moon itself would be cost effective and help to increase resource self-sufficiency. Research on the transformation of in situ raw materials is extremely important in that it will reduce the overall costs of exploration and settlement. Uppermost in the space traveler's mind is that in successfully tackling these lunar issues, many of the challenges that living on Mars present are also addressed.

If research, funding, and planning tasks can be spread among many interested parties who possess a variety of expertise and assets, the monumental job of getting settlements on the Moon can be accomplished. An ensemble of government entities, corporate or private companies, and individuals can cooperate to get these tasks completed. Collaboration with different motivations, skills, and resources is realistic and practical. For example, space tourism can pay for delivery of materials for a colony of permanent human habitations.

CONESTOGA IV
NASA

Transportation

The Moon, although not as big as the Earth, is big. There are no roads, rivers, or airports. Vehicles designed to reach locations for study or to acquire raw materials will be important. Transportation vehicles will need to function in a wide range of terrain: craters, mountains, steep slopes, and rocky landscapes. The vehicles need to provide life support, too. Both land and air crafts will be necessary.

Image set credit: NASA

Importance of Accessible Water

Finding accessible water on the Moon is important for further explorations and extended habitation. Water is needed for drinking, building, research, gardening for fresh food, and maintaining a general healthy life on the Moon. Water can be converted to oxygen for breathing. It can also be converted to both hydrogen and oxygen, which are important components of rocket fuel. More ancient sources of water are believed to be in the crust. More recent water sources on the lunar surface could be asteroidal in origin.

Image credit: NASA

Image credit: ISRO/NASA/JPL-Caltech/Brown Univ./USGS

Water at the Poles

In 2009, NASA's Moon Mineralogy Mapper, an instrument on the Indian Space Research Organization's Chandrayaan-1 mission, found water, which was confirmed by NASA. The water is located at the lunar poles and is illustrated by the blue colors. Also, in smaller quantities, water was detected at other locations.

Image set credit: NASA

Space Suits

NASA has developed specialized suits for different needs and environments. Astronaut Alan Shepard *(top left)* is in his silver pressure suit, prepared for his Mercury Redstone 3 launch on May 5, 1961. Astronaut Buzz Aldrin *(top right)* is on the Moon during the Apollo 11 mission with a suit that is made for a moonwalk rather than a spacewalk. Space shuttle Endeavour astronaut Dr. Mae Jemison *(middle left)* suits up for a launch. Training in Extravehicular Mobility Unit (EMU) spacesuits *(bottom left)* specialized for working in water are NASA astronauts Steve Swanson, Expedition 39 flight engineer and Expedition 40 commander, and Scott Tingle. Astronaut Peggy Whitsom, Expedition 16 commander, is suited for a spacewalk that lasted over seven hours.

Special Suits for Specific Purposes

Once astronauts are in a protected environment, the suits can come off and more normal clothing can be worn, but while they work and travel out of the spacecraft and living area, a space suit is necessary to stay alive. Space suits are designed to regulate temperature, pressure, water, and oxygen. The suit also provides light, communications, protection from radiation, and more. Improvements and innovations are always in development and will be specialized for work and life on the Moon.

Lunar and Mars Habitats

What we learn in building habitats on the Moon will be beneficial in designing and building Mars habitats. Many of the challenges are the same, although not identical. A breathable environment, lack of atmosphere, protection from the Sun and space weather, finding water, Moon and Mars quakes, local building materials, local resources for self sufficiency, and social isolation are just some of the difficult hurdles.

3D Printed, Crater, and Lava Tube Habitats

Occasionally, NASA will have competitions to research and solve problems. 3D-Printed Habitat Centennial Challenge was one of these competitions. The several top winners' images are included here *(left)*. Another variation is to print over an inflatable habitat module. The idea is to transform local materials, such as regolith,

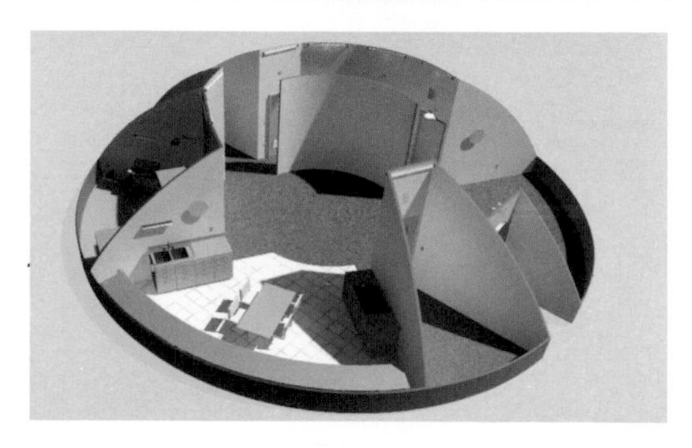

Image set credit: NASA

Image credit: esa

into livable structures. This 1.5 ton block *(top)*, on show in the laboratory corridor of ESA's ESTEC technical centre in the Netherlands, was 3D printed from material that simulated lunar dust.

This lunar crater *(right)* is about the size of the city of Philadelphia. It is being studied with the idea of covering it and using it as a protective environment for a settlement. These pit-type craters may give access to lava tubes lying beneath the surface and provide extensive, safe, and ready-made places for habitat building. There is speculation that the lava tubes could be sealed to create a breathable atmosphere.

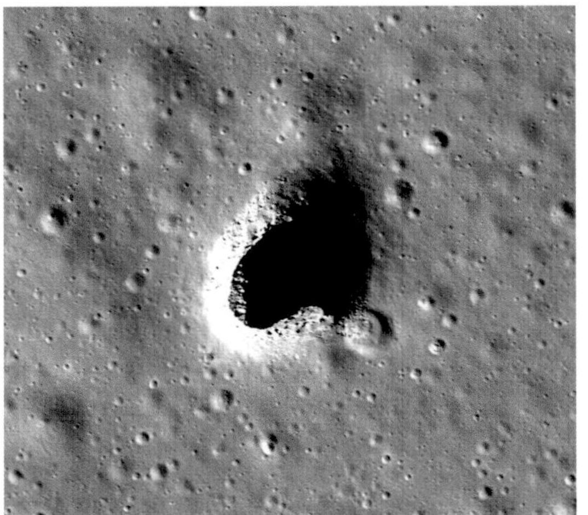
Image credit: NASA

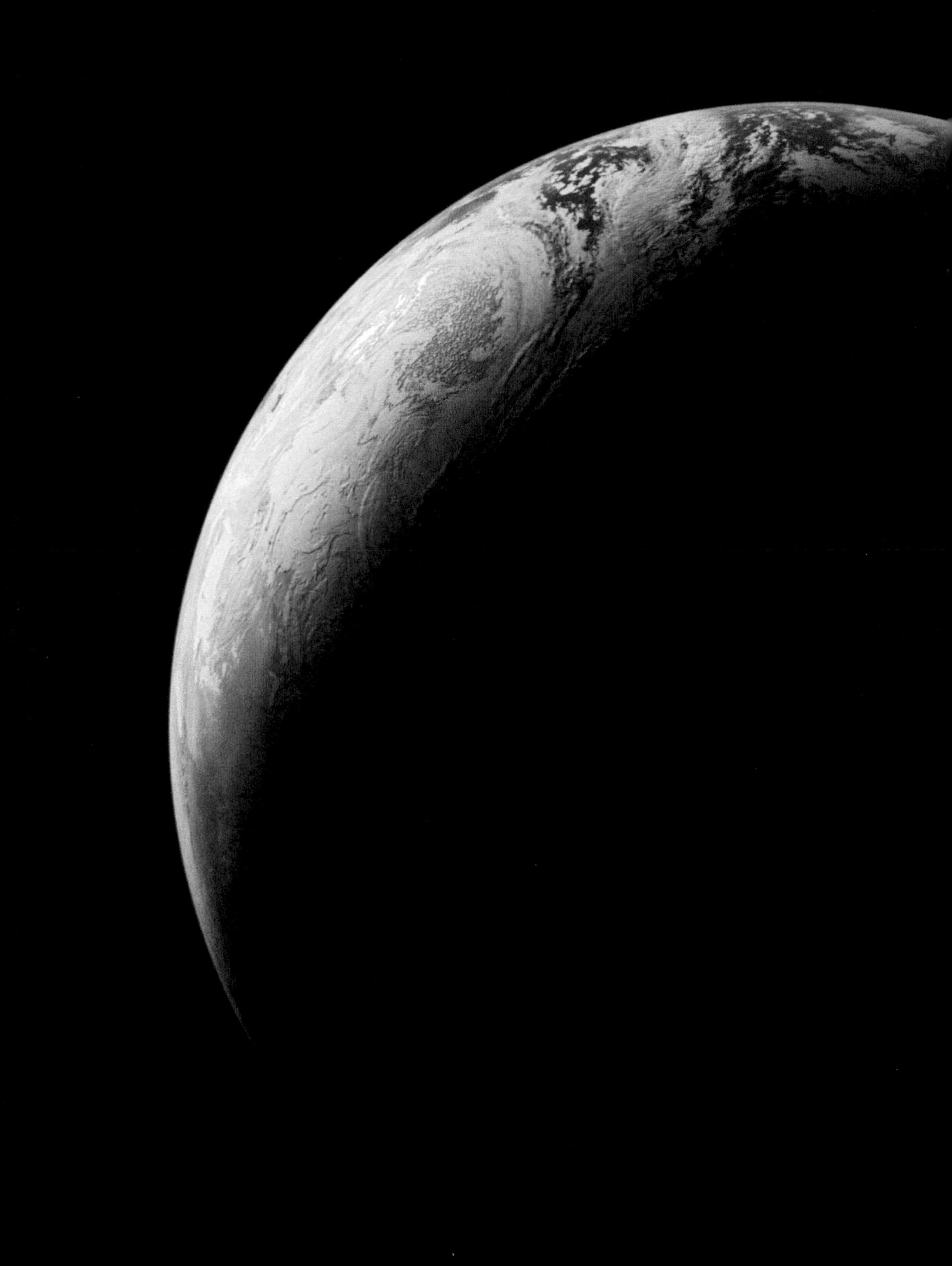

Earth Phases in the Moon's Sky

As the Moon orbits around the Earth, portions of the planet will appear lit by the Sun and other portions of the sphere will be in shadow. This image of the Earth *(left)* was taken on November 9, 1967 from the uncrewed Apollo 4-test flight. Another image taken during the Apollo 17 mission *(bottom)* shows a different phase. These phases are similar to what one would see from the Moon at different times of the month.

From the Moon, the Earth will appear to be stationary in the sky. However, in an Earth month (which is the same as a lunar day), the Earth will appear to go through all phases, similar to the Moon's phases: from a full disc, to a waning gibbous, to a third quarter, to a waning crescent, to a shadowed disc, to a waxing gibbous, to a third quarter, to a waxing crescent, and finally back to a full-Earth disc.

Image set credit: NASA

Earthrise

Living on the Moon, would we see an earthrise? The images on these pages suggest that we could. However, the image from NASA's Lunar Reconnaissance Orbiter (LRO) *(facing page)* is from the view of a moving spacecraft. Another apparent earthrise image *(top)* was taken by Bill Anders on Apollo 8 as the craft circumnavigated the Moon on December 24, 1968. The Earth appeared to rise and set because the spacecraft was going around the Moon.

Anyone viewing Earth from a static location on the Moon would always see the Earth stay in one place in the sky. It may wobble within that place and change in size because of its elliptical orbit. However, it will not rise and set.

The appearance of the Earth would change throughout the 24-hour day while remaining stationary in its sky. Since the Earth revolves once a day on its axis, the illuminated part of the ever-spinning Earth would present a changing face, repeating every 24 hours. This is unlike the Moon we see from Earth, which always presents the same side toward its planet.

Living on the Moon with this extraordinary difference in the sky may have a greater psychological effect than scientists anticipate. The experience of an ever-present Moon in a terrestrial sky's daily and monthly cycles would be replaced. The Sun's rising and setting would be the length of a month. The Earth would be doing a 24-hour spin constantly hanging in one place in the sky.

Image set credit: NASA

Future Lunar & Mars Missions

Gateway to the Moon

This is an infographic *(facing page)* in progress. The Gateway will be an international project. The blue modules are planned U.S. contributions. Purple modules are proposed international components, and yellow modules are joint U.S. and international or haven't been determined as of yet. The Gateway will be a small spaceship that will orbit the Moon. It will have living quarters, laboratories, offices, and docking ports. It will be a base for human and robotic missions to the Moon's surface. The finish date is projected to be in the year 2026.

The graphic below is a tentative timeline for NASA's lunar and Mars exploration campaign for approximately the next decade.

Image credit: NASA

GATEWAY CONFIGURATION CONCEPT

An exploration and science outpost in orbit around the Moon

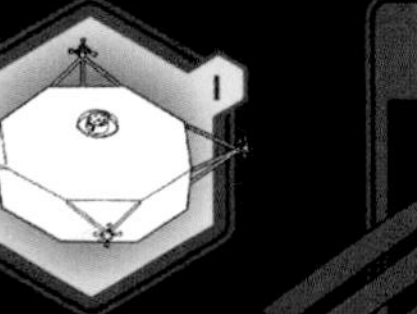

Power and Propulsion Element: Power, communications, attitude control, and orbit control and transfer capabilities for the Gateway.

ESPRIT: Science airlock, additional propellant storage with refueling, and advanced lunar telecommunications capabilities.

U.S. Utilization Module: Small pressurized volume for additional habitation capability.

Habitation Modules: Pressurized volumes with environmental control and life support, fire detection and suppression, water storage and distribution.

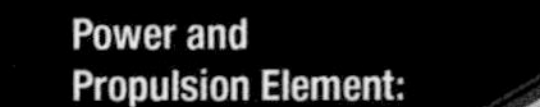
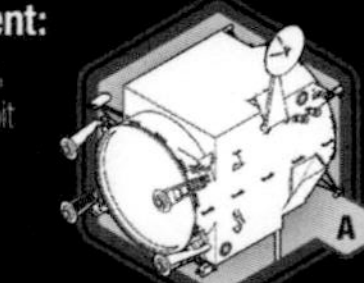
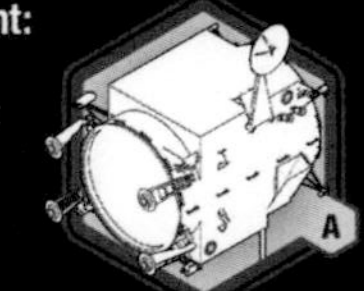

Robotic Arm: Mechanical arm to berth and inspect vehicles, install science payloads.

Logistics and Utilization: Cargo deliveries of consumables and equipment. Modules may double as additional utilization volume.

Airlock: Enables spacewalks, potential to accommodate docking elements.

Sample Return Vehicle: A robotic vehicle capable of delivering small samples or payloads from the lunar surface to the Gateway.

Orion: U.S. crew module with ESA service module that will take humans farther into deep space than ever before.

NASA-led architecture and integration

U.S.

International

TBD: U.S. and/or International

Gateway Compared to the International Space Station

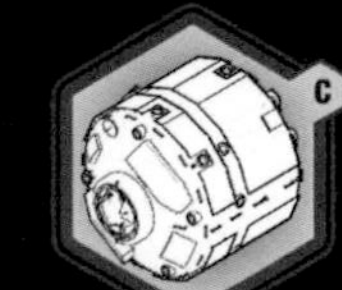
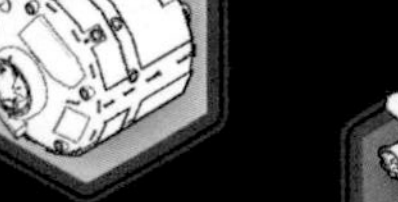

The International Space Station is a permanently crewed research platform that has 11 modules and is the size of a football field.

The Gateway is a much smaller, more focused platform for extending initial human activities into the area around the Moon.

Image credit: NASA

Reaching Mars Using a Close Approach

A day on Mars is nearly the same length as it is on Earth. In contrast, Earth's year is just under half as long as a Mars year. The two planets are found in different places in their orbits with respect to each other, and their distance apart changes. The two planets can be as close as 33.9 million miles (54.6 million kilometers) *(top)*, or if they are on opposite sides of the Sun *(bottom)* they can be as far as 240 million miles (401 million kilometers) away from each other. The term *close approach* refers to the time when two objects in orbit are at their closest. Planning and timing will be crucial for a crewed ship to reach Mars, and the Moon is pivotal in the overall strategy in preparing for and reaching Earth's closest neighbor using a close approach.

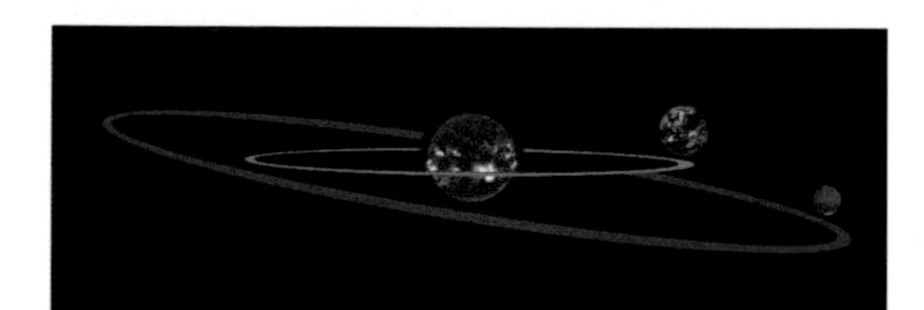

A Hop, Skip, and a Jump to Mars

We've received images from the uncrewed Mars Pathfinder *(facing page)* and current missions continue to study Mars. The Moon and its orbit are the ideal training ground for crewed missions to Mars to learn self-reliance skills, to refine necessary technologies, to stockpile provisions, and developing away-from-Earth strategies.

Image set credit: NASA

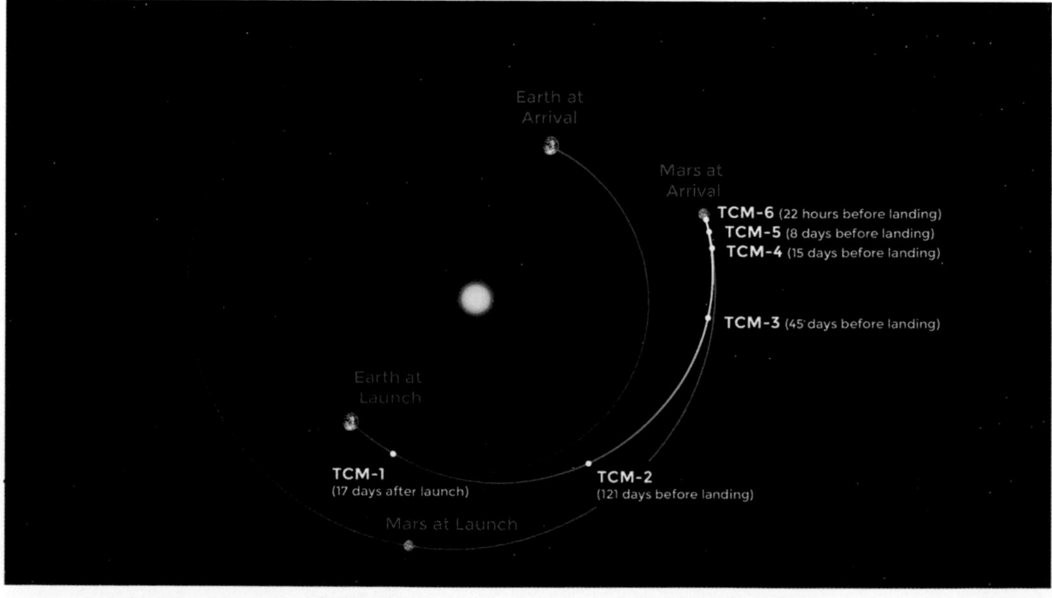

The Moon's Future

The future of the Moon depends on humanity's ability to work collectively among nations, private enterprises, and key individuals. The combined goodwill of the human race is the foundation for our successful and beneficial presence in space. And we are well on our way. A series of large tasks is required to create the international Gateway to the Moon and explore beyond to other planets. Scientific breakthroughs that help get the job done will abound in this work, but will have many applications for improving life on Earth too. Humanity benefits from the scientific, prepared and methodical, the innovative, courageous, spirited, curious, and pioneering individuals who participate in this exploration of the Moon.

Once, the vast and challenging distance across oceans—between Earth's continents—were traversed for the first time. Now, it is the vast and challenging distance between the Moon and planets that lie before us. Whether individuals leave footprints in the

Image set credits: NASA/JPL-Caltech

Moon's dust or operate a keyboard from the Earth, humanity is about to become more intimate with the Moon. It's only a matter of a short time before it is an established, populated, extension of the civilized world. And the Moon, like an off-world continent, will be a stopping point for other destinations.

Index